Report of Investigations 9680

Evaluation of Face Dust Concentrations at Mines Using Deep-Cutting Practices

J. Drew Potts, W.R. Reed, and Jay F. Colinet

DEPARTMENT OF HEALTH AND HUMAN SERVICES
Centers for Disease Control and Prevention
National Institute for Occupational Safety and Health
Office of Mine Safety and Health Research
Pittsburgh, PA • Spokane, WA

January 2011

This document is in the public domain and may be freely copied or reprinted.

Disclaimer

Mention of any company or product does not constitute endorsement by the National Institute for Occupational Safety and Health (NIOSH). In addition, citations to Web sites external to NIOSH do not constitute NIOSH endorsement of the sponsoring organizations or their programs or products. Furthermore, NIOSH is not responsible for the content of these Web sites.

The findings and conclusions in this report are those of the author(s) and do not necessarily represent the views of the National Institute for Occupational Safety and Health.

Ordering Information

To receive documents or other information about occupational safety and health topics, contact NIOSH at

> Telephone: **1–800–CDC–INFO** (1–800–232–4636)
> TTY: 1–888–232–6348
> e-mail: cdcinfo@cdc.gov
>
> or visit the NIOSH Web site at **www.cdc.gov/niosh**.

For a monthly update on news at NIOSH, subscribe to NIOSH *eNews* by visiting **www.cdc.gov/niosh/eNews**.

DHHS (NIOSH) Publication No. 2011–131

January 2011

SAFER • HEALTHIER • PEOPLE™

Table of Contents

Abstract 1
Introduction 2
Sampling Protocol and Data Analysis 3
Underground Dust Surveys 6
Exhausting Ventilation 7
Blowing Ventilation 9
No-curtain Ventilation 11
Bolting Operations 12
Recent Developments Using Blocking Sprays 13
Conclusions 14
References 15
Appendix A: MINE A CASE STUDY 25
    Mine-specific Information 25
    Shuttle Car Data Analysis 26
    Miner-generated Dust 29
    Miner Operator Data Analysis 29
    Bolter Operations Data Analysis 30
    Dust-control Monitoring 30
    Summary 31
Appendix B: MINE B CASE STUDY 37
    Mine-specific Information 37
    Shuttle Car Data Analysis 38
    Miner-generated Dust 40
    Miner Operator Data Analysis 41
    Bolter Operations Data Analysis 41
    Summary 42
Appendix C: MINE C CASE STUDY 46
    Mine-specific Information 46
    Shuttle Car Data Analysis 47
    Miner-generated Dust 48

Miner Operator Data Analysis 49
Bolter Operations Data Analysis 49
Dust-control Monitoring 50
Summary 50

Appendix D: MINE D CASE STUDY 57

Mine-specific Information 57
Shuttle Car Data Analysis 58
Miner-generated Dust 61
Miner Operator Data Analysis 62
Bolter Operations Data Analysis 62
Dust-control Monitoring 63
Summary 63
References Cited in Appendix D 64

Appendix E: MINE E CASE STUDY 69

Mine-specific Information 69
Shuttle Car Data Analysis 71
Miner-generated Dust 75
Bolter Operations Data Analysis 76
Dust-control Monitoring 76
Summary 77
References Cited in Appendix E 78

Appendix F: MINE F CASE STUDY 83

Mine-specific Information 83
Shuttle Car Data Analysis 85
Miner-generated Dust 86
Miner Operator Data Analysis 86
Bolter Operations Data Analysis 87
Dust-control Monitoring 88
Summary 88
Reference Cited in Appendix F 89

Figures

Figure 1. Area sampling locations for a typical section layout. Note: PDMs Are not shown. 17

Figure 2. Pie chart depicting the magnitude of scrubber airflow reduction during the cut as a percentage of total cuts measured. 21

Figure 3. Mine operator and shuttle car cab position at the start of the no-curtain cut for the initial heading development beyond the last open crosscut. 22

Figure 4. Mine operator and shuttle car cab position at the end of the no-curtain cut for the initial heading development beyond the last open crosscut. 23

Figure A-1. Mine A continuous miner spray configuration. 25

Figure A-2. Plan view of the mine A operating section. 26

Figure A-3. Shuttle car dust exposures with exhausting face ventilation when the miner was driving a heading. 27

Figure A-4. Shuttle car dust exposures with blowing face ventilation when the miner was driving a heading. 28

Figure A-5. Shuttle car dust exposures with blowing face ventilation when the miner was turning a right crosscut. 29

Figure B-1. Mine B continuous miner spray configuration. 37

Figure B-2. Plan view of Mine B section. 38

Figure B-3. Shuttle car dust levels when miner was driving a straight heading. 39

Figure B-4. Miner return dust levels during Cut No. 4. 41

Figure C-1. Mine C continuous miner spray configuration. 46

Figure C-2. Plan view of the mine C continuous miner development section. 47

Figure D-1. Mine D continuous miner spray configuration. 57

Figure D-2. Plan view of mine D continuous miner section. 58

Figure D-3. Shuttle car dust levels at the face when the miner was driving headings without the use of line curtain. 60

Figure D-4. Shuttle car dust levels at the face when the miner was driving right crosscuts without the use of curtain. 60

Figure D-5. Shuttle car dust levels at the face when the miner was driving a heading with the use of blowing curtain. 61

Figure D-6. Miner return dust levels at 5-sec intervals for each cut. 62

Figure E-1. Mine E continuous miner spray configuration. 69

Figure E-2. Plan view of the mine E continuous miner Section No. 4. 70

Figure E-3. Plan view of the satellite entries driven off the mains on Section No. 4. ... 70

Figure E-4. Shuttle car cab dust levels at the face when driving a heading with the use of blowing curtain. 73

Figure E-5. Shuttle car cab dust levels at the face when driving an entry without the use of blowing curtain. 73

Figure E-6. Shuttle car cab dust levels at the face when turning a right crosscut with the use of blowing curtain. 74

Figure E-7. Shuttle car cab dust levels at the face when turning a left crosscut with the use of blowing curtain. 74

Figure E-8. Miner return dust levels at 5-sec intervals for each cut that did not use ventilation curtain. 76

Figure F-1. Mine F continuous miner spray configuration. 84
Figure F-2. Plan view of the operating section at Mine F. 85
Figure F-3. Shuttle car cab dust levels when loading at the active face. 86

Tables

Table 1. Shuttle car cab dust levels when loading at the face 18
Table 2. Time weighted average levels for the shuttle car, miner, and bolter operators 19
Table 3. Miner operator exposure levels and miner-generated 20
dust levels during regular- and deep-cuts depths 20
Table 4. Bolter operator exposure levels and machine-generated 24
dust levels at regular- and deep-cut depths 24
Table A-1. Shuttle car dust levels when loading at the face for Cuts No. 1, 2, and 3 33
Table A-2. Shuttle car dust levels when loading at the face for Cuts No. 4, 5, and 6 34
Table A-3. Shuttle car dust levels when loading at the face for Cut No. 7 35
Table A-4. Continuous miner-generated dust during production as measured with pDRs calibrated with gravimetric samplers 35
Table A-5. Miner operator exposures during production as measured with a PDM instrument 36
Table A-6. Bolter-generated dust and bolter operator exposures during the bolting cycle 36
Table B-1. Shuttle car dust levels when loading at the face during Cuts No. 1 through 3 43
Table B-2. Shuttle car dust levels when loading at the face during Cuts No. 4 through 6 44
Table B-3. Miner-generated dust levels during the regular- and deep-cut depths for each cut 45
Table B-4. Miner operator dust levels during the regular- and deep-cut depths for each cut 45
Table B-5. Bolter operator exposure levels during the regular- and deep-cut depths for each bolted room 45
Table C- 1. Shuttle car dust levels when loading at the face during Day 1 of the study 52
Table C- 2. Shuttle car dust levels when loading at the face during Day 2 of the study 53
Table C- 3. Shuttle car dust levels when loading at the face during Day 3 of the study 54
Table C- 4. Miner-generated dust levels during the regular- and deep-cut depths 55
Table C- 5. Miner operator dust exposure levels net of intake dust level during the regular- and deep-cut depths 55
Table C- 6. Bolter machine-generated dust levels and bolter operator exposures for the regular- and deep-cut depths 56
Table D-1. Shuttle car cab dust levels when loading at the face during Day 1 of the study 65
Table D-2. Shuttle car cab dust levels when loading at the face during Day 2 of the study 66

Table D-3. Shuttle car cab dust levels when loading at the face during Day 3 of the study 67
Table D-4. Miner-generated dust levels during the regular- and deep-cut depths 68
Table D-5. Bolter-generated dust levels and operator exposures for each bolting sequence 68
Table E-1. Shuttle car cab dust levels when loading at the face during Cuts No. 1 through 4 79
Table E-2. Shuttle car cab dust levels when loading at the face during Cuts No. 5 through 7 80
Table E-3. Shuttle car cab dust levels when loading at the face during Cuts No. 8 through 10 81
Table E-4. Miner-generated dust levels during the deep- and regular-cut depths 82
Table E-5. Bolter-generated dust levels and operator exposures for each bolting sequence 82
Table F-1. Shuttle car cab dust levels when loading at the face on Day 1 90
Table F-2. Shuttle car cab dust levels when loading at the face on Day 2 91
Table F-3. Shuttle car cab dust levels when loading at the face on Day 3 92
Table F-4. Miner-generated dust levels during the deep- and regular-cut depth 93
Table F-5. Miner operator dust levels during the deep- and regular-cut depths 93
Table F-6. Bolter-generated dust levels and operator exposures for each bolting sequence 94

ACRONYMS AND ABBREVIATIONS

CFR	Code of Federal Regulations
CI	confidence interval
EIA	Energy Information Administration
MSHA	Mine Safety and Health Administration
NIOSH	National Institute for Occupational Safety and Health
No.	number
PDM	personal dust monitor
pDR	personal DataRAM
PVC	polyvinyl chloride
RAM	real-time aerosol monitor
TWA	time weighted average
USBM	United States Bureau of Mines

UNIT OF MEASURE ABBREVIATIONS USED IN THIS REPORT

cfm	cubic foot per minute
°	degree
ft	foot
fpm	foot per minute
gpm	gallon per minute
in	inch
in Hg	inch of mercury
µg/m^3	microgram per cubic meter
mg/m^3	milligram per cubic meter
mm	millimeter
%	percent
psi	pound-force per square inch

Evaluation of Face Dust Concentrations at Mines Using Deep-Cutting Practices

J. Drew Potts, W.R. Reed, and Jay F. Colinet

National Institute for Occupational Safety and Health

Abstract

Dust surveys were conducted at six underground mines to determine if deep-cut mining practices expose face workers to higher levels of respirable dust by comparing levels during the first 20 ft of advance (regular-cut depth) during the deep cut to levels during the final 10 to 20 ft of advance (deep-cut depth). The studies were conducted at mines where the Mine Safety and Health Administration (MSHA) had approved an extended curtain setback distance with operation of a flooded-bed scrubber to permit taking deep cuts of up to 40 ft. In general, all of the selected mines exercised good dust control practices by maintaining water sprays, scrubber airflows, proper curtain setback distances and providing sufficient airflow to the active faces. These practices minimized variability in dust levels related to factors other than the depth of cut. To ensure proper scrubber functioning, the scrubber screen was back-flushed before commencing each cut. Both exhausting and blowing face ventilation configurations were studied. All of the operations surveyed for this study were able to successfully implement deep-cutting methods without significantly increasing the dust exposures of face workers during the cutting and bolting cycles.

For exhausting face ventilation, field data indicate that scrubber airflow is the most important factor for controlling dust. Clogging of the scrubber screen can result in lower airflows; therefore, the screen must be periodically tapped and back-flushed. Data collected for this study indicate that 20-mesh screens should be cleaned for every 40 ft of advance because 22% of the deep-cut sequences surveyed for this study experienced a 20% to 35% decrease in scrubber airflow over the course of the cut. For blowing face ventilation, field and laboratory data indicate that maintaining a proper curtain-to-scrubber airflow ratio of 1.0 and a curtain setback distance that allows the miner operator to stand at the mouth of the curtain helps control dust. Curtain airflows should be measured before activation of the scrubber regardless of ventilation type (exhausting or blowing) to avoid erroneously overinflating the ratio. The curtain setback variance should be greater than the maximum cutting depth to allow miner operators to maintain their position at the mouth of the curtain when the miner is fully extended into the cut. Greater curtain setback distances associated with deep-cutting methods may result in cuts that do not require ventilation curtain, such as the initial heading developments beyond the last open crosscut. For these cuts, dust levels were generally lower during development of deep cuts when compared to regular cuts. However, adequate ventilation of cuts without ventilation curtain is dependent on a properly functioning scrubber. Dust levels on the bolting faces did not appear to be affected by the longer cycles associated with deep-cut mining practices when curtain airflow was measureable and the curtain was periodically advanced in sync with the bolting machine.

Introduction

In 2008, underground coal mines in the United States produced over 357 million tons of coal [EIA 2008], with over half of this production coming from continuous mining operations. MSHA determined that over 800 continuous mining units were in operation in 2010 [Niewiadomski 2010]. During the mining of coal, respirable coal dust is liberated into the air, potentially exposing mine workers. Likewise, during the extraction of the coal, rock within or surrounding the coal seam may be cut, liberating respirable silica into the air. Overexposure to respirable coal dust can lead to the development of coal workers' pneumoconiosis, while overexposure to silica dust can lead to silicosis. Both of these lung diseases can be disabling or fatal, depending upon the severity of the disease that develops. Consequently, to prevent the development of occupational lung disease, mining operations are responsible for controlling concentrations of respirable dust in the mine atmosphere where miners work or travel.

The Federal Coal Mine Health and Safety Act of 1969 (Public Law 91-173), generally referred to as the Coal Act, established the current respirable coal mine dust standard of 2.0 mg/m^3. Compliance with this standard is monitored through periodic collection of occupational and area dust samples with an approved gravimetric sampling pump. However, if the silica content of the collected sample is greater than 5%, then a reduced dust standard is calculated by dividing 10 by the percent silica in the sample. For example, if the compliance sample contains 10% silica, a reduced dust standard of 1.0 mg/m^3 is enforced on that mining unit. Enforcement of a reduced dust standard is designed to limit exposure to respirable silica dust to a maximum of 100 µg/m^3.

The Mine Safety and Health Administration (MSHA) is responsible for enforcing the respirable dust regulations. Prior to starting a new mining section, mine operators are required to develop a mine ventilation plan designed to control methane and respirable dust, and they must submit the plan to MSHA for approval. The dust control portion of the mine ventilation plan specifies the types of dust controls, work practices, and maintenance procedures to be used on each shift to limit dust concentrations to acceptable levels.

Implementation of these dust regulations has had a significant impact on the respirable dust exposure of mine workers and on the prevalence of lung disease. In April of 1968, the U.S. Bureau of Mines (USBM) conducted sampling at over 20 mines to quantify respirable dust levels for select occupations [USBM 1970]. Results of this sampling showed that the average dust exposure levels were 5.63 mg/m^3 for the continuous miner operator and 3.39 mg/m^3 for the roof bolter operators [Colinet et al. 2010]. For MSHA inspector samples collected in 2009, the average dust exposure levels were 0.81 mg/m^3 for the continuous miner operator and 0.60 mg/m^3 for roof bolter operators. These average dust exposure levels represent all valid samples collected for these occupations and show substantial reductions compared to the 1968 results. Unfortunately, samples show that individual shift concentrations above the applicable standards continue to occur. For example, continuous miner operator samples from 2006 through 2009 show that 6.7% of the samples exceeded 2.0 mg/m^3, while 13.0% of samples exceeded their reduced dust standard [MSHA 2010].

Traditionally, ventilation and water sprays have been the primary methods used to control dust generation from the continuous miner. The miner was operated by a worker positioned in an operator's cab located at the back right corner of the machine. This worker was not permitted to travel under unsupported roof, thereby limiting the depth of mining cuts to 20 ft.

As continuous mining equipment was modified, new mining techniques also evolved. Today, most continuous mining machines are equipped with remote control units and flooded-bed scrubbers. The remote control unit has allowed the miner operator to move off of the miner and stand in the entry while operating the machine. Therefore, the miner operator can remain under supported roof while the mining machine advances under unsupported roof to extract cuts that extend beyond 20 ft. However, for MSHA to approve curtain setback distances greater than 20 ft to permit deep cuts, the mine must demonstrate that the roof integrity will allow for advances beyond 20 ft and adequate ventilation can be provided to the face to control methane and dust levels. The flooded-bed scrubber is a fan-powered dust collector, which induces airflow to the face and collects over 90% of the respirable dust drawn into the unit when properly maintained [McClelland et al. 1992]. Consequently, the combination of remote control and a flooded-bed scrubber has allowed continuous miner units to extract deep cuts. Approximately 70% of miner units are extracting deep cuts [Niewiadomski 2010].

On June 3, 2008, MSHA issued a procedure instruction letter entitled *Procedures for Evaluation of Requests to Make Extended Cuts With Remote Controlled Continuous Mining Machines* (PIL No. I08-V-3). This procedure provides guidance for evaluating requests for approval of deep cuts and raises emphasis on effectively controlling respirable dust throughout the entire cut. Stakeholders inquired about available data regarding dust levels for different portions of deep cuts, but NIOSH was not aware of any existing data. In response, NIOSH initiated an underground study to evaluate dust levels generated throughout the extended cut, while also monitoring the performance of the dust collectors used on the miners and roof bolters. NIOSH used sampling instrumentation and techniques to allow for a comparison of dust levels generated during advancement of the first 20 ft (regular cut) of a deep cut to levels during advancement of the final 10 to 20 ft (deep cut) of a deep-cut. Dust surveys were conducted at six mines under a variety of operating conditions, with the results summarized in this report.

Sampling Protocol and Data Analysis

The dust sampling protocol used both area and personal sampling techniques to evaluate face dust concentrations on continuous mining sections using deep-cutting methods. Neither the area nor personal sampling techniques used to conduct the underground surveys followed compliance sampling requirements prescribed by Title 30 of the Code of Federal Regulations (CFR). More specifically, the sampling devices were not operated portal to portal (30 CFR 70.201), nor were they converted to an equivalent concentration as measured with an MRE instrument (30 CFR 70.206). Rather, all of the concentrations reported are based on time weighted averages (TWAs). Area samplers were turned on and off at the working faces, and sampling times average between five and seven hours. The personal sampling devices were programmed to start operating when the workers reached the face, and to run for a period of eight hours. Respirable dust measurements were made using the following real-time and gravimetric sampling equipment:

(1) Thermo Scientific Model 1000AN personal DataRams (pDRs)
(2) Thermo Scientific Model 3600 personal Dust Monitors (PDMs)
(3) MSA Escort Elf sampling pumps with MSA filters and Dorr-Oliver 10-mm cyclones (gravimetric samplers)

Figure 1 shows a plan view of the area sampling packages for a section layout when a exhausting curtain was used to ventilate the active face. Area sampling was conducted with a package consisting of two gravimetric samplers and one pDR. The pDR utilizes light scattering technology to measure relative dust levels in real-time, which are stored in an internal data logger. The data collected with the pDR, combined with a detailed time study, allowed researchers to determine dust concentrations during specific periods of time. The gravimetric samples were used to calibrate the pDRs and consisted of MSA pumps that drew dust-laden air at 2.0 liters/minute through 10-mm nylon cyclones to deposit the respirable dust fraction onto pre-weighed 37-mm PVC filters. All filters were pre- and post-weighed at the NIOSH controlled environment weighing lab in Pittsburgh, providing the data from which respirable dust concentrations were calculated. Correction factors to calibrate the pDR data, as recommended by the manufacturer (Thermo Scientific), were calculated by dividing the daily average gravimetric concentration by the daily average pDR concentration for each sampling location. The instantaneous readings from the pDRs were then multiplied by the correction factors. Isolation of bolter- and miner-generated dust was accomplished by placing the area sampling packages in the immediate intake and return for each piece of equipment. Intake dust levels were subtracted from return dust levels to determine the portion of dust attributable to a particular piece of equipment. Area sampling packages were also hung in the shuttle car cabs to monitor potential operator exposure levels.

Due to the mobility of the bolter and miner operators, belt-wearable PDMs were used, instead of area sampling packages, to monitor the dust exposures of these workers. PDMs are mass-based samplers, which provide real-time dust measurements that do not require a gravimetric calibration.

Real-time dust data allowed researchers to analyze dust level fluctuations from start to finish for each deep cut. A detailed time study was conducted to record when each shuttle car entered and exited the active face. The shuttle car cab dust levels were of particular importance for this study because each car's position with respect to the mining machine remained consistent throughout advancement of the face. This data allowed direct determination of dust levels around the mining machine at various depths. Of particular importance was a comparison of dust levels during the first 20 ft of advancement, which represents a regular-cut depth, to dust levels during the last 10 to 20 ft of advancement, the deep-cut depth. Miner operator exposure levels and miner-generated dust levels were also determined for the regular- and deep-cut depths. Mines using deep-cutting methods also have longer bolting cycles. Therefore, a similar time study was conducted on the bolter faces, allowing a comparison of dust level fluctuations as the bolting machine advanced.

Since the objective of the study was to compare dust levels during the regular cut to levels during the deep cut, it was important to examine variables that changed during an individual cut sequence. During our field studies we were able to measure two operating parameters, productivity and scrubber airflow, that changed frequently when comparing the regular cut to the deep cut. It is a generally accepted principle and common practice to assume that dust levels and productivity exhibit a linear relationship with one another. For example, if cut A was mined at a production rate of 1 car per minute and cut B was mined at a production rate of 2 cars per minute, one would expect cut B concentrations to be double that of cut A. Therefore, all of the miner operator exposure levels and mining machine-generated dust levels presented in this report were normalized based on the measured productivity for each cut sequence. The number of cars loaded per minute was used as the measure of productivity, and dust levels for each particular mine were normalized to the average productivity observed throughout the survey. For example, if the average productivity for the section was 0.37 cars per minute and productivity during the regular-cut depth for Cut No. 1 was 0.41 cars per minute, then the dust levels measured for this cut were normalized by multiplying by a correction factor of 0.90 (0.37/0.41).

Data analysis focused on determining whether or not dust levels during the deep-cut depth were significantly different than dust levels during the regular-cut depth. When determining statistical significance, the most commonly used confidence level for scientific experimentation is 95%. A less stringent but still robust standard that involved calculating 85% confidence intervals (CIs) for the sample means and comparing the values for overlap was used for this project due to the nature of the collected data. It is often difficult to obtain enough underground data to document changes in dust levels using a 95% confidence level criterion because fluctuations in operating parameters that are known to affect these levels can occur within and between individual cuts, adding variability to the data. These operating parameters include, but are not limited to, curtain airflows and setback distances, water spray pressures and quantities, and positioning of mine personnel. Using the 85% CI calculations allowed researchers to draw conclusions from the data with a reasonable degree of certainty. Comparisons that met more stringent statistical significance criterion are noted in the text.

Measurements other than dust were necessary to fully understand the effects of deep-cutting methods on face dust levels. As the miner advances into the deep cut, a buildup of material may occur in the flooded-bed scrubber filter, diminishing its effectiveness. This situation was monitored by measuring scrubber airflow at the beginning and end of each cut using a pitot tube and manometer. At the beginning of the study, velocity head traverse readings of the scrubber duct were measured using a pitot tube and manometer. The traverse readings included top, center, and bottom measurement points at each sampling port provided by the manufacturer, resulting in 15 to 18 readings. Scrubber airflow was determined based on the average of the readings. Succeeding scrubber airflow measurements were made using a single-point pitot tube reading, which was calibrated to the traverse readings.

Likewise, with the bolting machine, a buildup of material in the dust collection box may diminish the performance of the dry dust collection system. Therefore, the suction pressure of the bolter's dust collection circuit was periodically measured and recorded throughout the bolting sequence, using a vacuum gauge at the drill heads.

Airflow delivered to the active cutting and bolting faces was measured using an electronic vane anemometer. Measurements were taken at the face-side mouth of the curtain. Following mine protocol, curtain airflow at the exhausting face ventilation mines was measured before activation of the scrubber and after activation at the blowing face mines. Curtain configuration and length were also recorded.

Underground Dust Surveys

A total of six underground dust surveys were conducted to complete this project. Two surveys were conducted at mines using exhausting face ventilation, two at mines using blowing ventilation, and two at mines using either blowing or exhausting curtain, depending on their location on the section. In-depth mine case studies on the findings at each of these mines, labeled A through F, are included in the Appendices. All of the mines were able to use deep-cutting practices without hindering their ability to limit the dust exposures of face workers. However, researchers were able to identify some practical operating parameters to ensure the successful use of deep-cutting methods from the perspective of dust control.

Table 1 shows a summary of the 47 cuts examined for this study. The table includes face ventilation type, orientation of the cut, a mine identifier (A through F), curtain airflow, the starting and ending scrubber readings, curtain-to-scrubber airflow ratios, and shuttle car cab dust levels when loading at the face during the regular and deep portions of the cut. The number of data points used to calculate each reported shuttle car cab dust level in Table 1 ranged between 6 and 13, with the average equal to 9. These numbers are derived from the data tables in the appendices. Table 1 also shows a determination as to whether or not the differences in dust levels between the deep and regular cuts were significant by comparing the 85% CIs for the averages. Following mine-specific protocol, the curtain airflows presented in Table 1 were measured before activation of the scrubber for the exhausting faces and after activation of the scrubber for the blowing faces. Combining all data presented in Table 1, the average shuttle car cab dust levels when loading at the face were 0.96 mg/m^3 during regular-cut depth and 1.04 mg/m^3 during the deep-cut depth, which was not a statistically significant (85% CIs) difference. As mentioned earlier, shuttle car cab dust levels when loading at the face were used as the primary indicator of face area dust levels because their positions with respect to the mining machine remained consistent throughout advancement of the face.

Some of the shuttle car cab dust levels that are presented in Table 1 for the blowing face ventilation cuts appear to be relatively high. However, the levels represent short-term exposures during cutting and loading activities and are not reflective of full-shift exposures. Most of the shuttle cars' time was spent in transit to and from the feeder/breaker. For example, the highest shuttle car dust levels when loading at the face were measured at Mine A when using blowing face ventilation. The average value of these levels (from Appendix A, Tables A-1, A-2, and A-3) equaled 4.84 mg/m^3. When loading, transit, and the dumping sequence were included in the calculation at Mine A, the average shuttle car cab dust level during blowing face ventilation cuts was 1.98 mg/m^3. Furthermore, when miner down and move times were included in the calculation of the shuttle car cab dust level, the TWA exposure for the entire sampling period was 1.20 mg/m^3 on day 2 of the study at Mine A when all of the cuts were made using blowing face ventilation.

Table 2 shows all of TWA exposure levels for the entire sampling periods at the miner, bolter, and shuttle car operators for each of the surveyed mines. Two miner operator exposures are listed for Mine C because two mine workers alternated between operator and helper duties on an every-other-cut basis. The missing data in Table 2 are explained in the Appendices. The highest

TWA exposure level measured during the study was 1.76 mg/m^3, and most levels were below 1.0 mg/m^3.

Table 3 shows miner operator exposure levels and miner-generated dust levels during the regular- and deep-cut sequences for each of the 47 cuts listed in Table 1. All of the presented values were normalized for productivity. Combining all data, the average miner operator exposures were 0.95 mg/m^3 during the regular-cut depth and 0.93 mg/m^3 during the deep-cut depth. The average miner-generated levels were 2.02 mg/m^3 during the regular-cut depth and 1.23 mg/m^3 during the deep-cut depth. Only the difference in miner-generated dust levels was significant (85% CIs). This significant difference occurred due to the no-curtain cuts, which is explained in the "No-curtain Ventilation" section of this report.

An analysis follows for the different types of face ventilations encountered during the study, including exhausting, blowing, and no-curtain. These sections of the publication draw information from the Appendices and present an inter-mine perspective in order to draw conclusions regarding the use of the deep-cutting method and its effects on respirable dust levels in the bolting and mining face areas.

Exhausting Ventilation

In general, face area dust levels, as measured in the shuttle car cabs when loading at the face, were quite low during the exhausting face ventilation cuts conducted at Mines A, B, and C, averaging 0.20 mg/m^3 during regular-cut depth and 0.35 mg/m^3 during deep-cut depth. This difference is not statistically significant (85% CIs).

A significant difference was measured for several cuts at Mine C. This mine used two shuttle cars to transport coal to the feeder/breaker. One car had a standard cab configuration, while the other was equipped with an off-standard cab configuration. The analysis completed for Mine C, which is detailed in Appendix C, found that dust levels in the off-standard cab were 0.2 to 0.4 mg/m^3 higher (95% CIs) during the deep-cut sequence when compared to the regular-cut sequence. This difference may have resulted because the off-standard cab was on the same side of the entry as the exhaust tubing and, as the cut was advanced, the cab's position approached the mouth of the tubing, where it was exposed to dust rollback. This difference was not observed in the standard cab, which was on the off-curtain side of the entry. A laboratory study conducted by NIOSH made similar observations concerning the off-standard side cab. The study found that increasing the curtain setback distance from 30 to 40 ft, significantly increased dust levels at the off-standard shuttle car position by placing it near the mouth of the curtain [Goodman et al. 2006]. Therefore, when using deep-cutting methods with exhausting face ventilation, it is beneficial to use shuttle cars with cabs located on the off-curtain side of the entry to help ensure that dust levels do not increase with increasing depth of cut.

A stepwise linear regression analysis was conducted on the exhausting face ventilation data contained in Table 1 to determine if independent variables affected shuttle car cab dust levels (dependent variable), including curtain airflow, scrubber airflow, and the curtain-to-scrubber airflow ratio. The mine identifier was also included as an independent variable to account for any inter-mine differences. Calculations of the values for kurtosis (2.1) and skewness (1.5) associated with the dependent variable did not indicate a substantial deviation from an assumption of normality. Other model assumptions included the following:

(1) Curtain airflow remained constant throughout the cut.
(2) The starting scrubber airflow prevailed during the regular-cut depth.
(3) The ending scrubber airflow prevailed during the deep-cut depth.

The scrubber airflows presented in Table 1 for Mine C were the only levels measured and reported by mine personnel and based on single center-point pitot tube readings. Since center-point readings are typically higher than full-traverse calculations, actual scrubber levels for Mine C were probably somewhat lower than those reported in the table. All other scrubber airflow measurements were conducted by NIOSH personnel and were adjusted based on a full-traverse pitot tube study of the scrubber duct as explained in the "Sampling Protocol and Data Analysis" section of this report.

Scrubber airflow was the only independent variable found to be significant (95% CIs) for the stepwise linear regression analysis. Due to the presence of collinearity between the scrubber and curtain airflows, a cause-and-effect relationship between curtain airflow and dust levels cannot be ruled out, however, scrubber airflow explained more of the variability in the dust levels. Dust levels increased as the scrubber airflow decreased for cuts using exhausting face ventilation. The importance of scrubber airflow when using exhausting face ventilation was also identified in a recent laboratory study, which concluded that a 20% decrease in scrubber airflow produced a significant increase in dust levels on the curtain side of the machine as well as in the return [Organiscak and Beck, forthcoming]. As shown in Figure 2, a pie chart depicting the reductions in scrubber airflows measured from start to finish for all of the 47 deep cuts surveyed for this study, 22% of the cuts experienced a drop in the range of 20% to 35%. This figure conveys the importance of keeping the scrubber screens clean. All of the mines in the study used 20-mesh screens and experienced some average reduction in scrubber airflow during the course of the deep cut. Typical mine practices, as prescribed by the mine ventilation plans and manufacturer recommendations, involve cleaning the scrubber screens at least twice during a shift. The above data suggest that cleaning the screens by tapping out the contaminants and back-flushing with water should be conducted before each deep cut to ensure maximum scrubber airflow.

For the exhausting ventilation cuts shown in Table 3, the average miner operator exposure levels during the regular- and deep-cut depths were both low at 0.55 mg/m^3 during the regular-cut depth and 0.83 mg/m^3 during the deep-cut depth. This difference is not statistically significant (85% CIs). In general, all of the exhausting face ventilation mines surveyed for this study positioned the miner operator on the off-curtain side of the entry in an area that was parallel to, or outby, the inlet mouth of the curtain. From a dust control perspective, research has shown this to be the best location for the operator [Colinet and Jankowski 1996]. Curtain setback distances were maintained between 30 and 50 ft at the surveyed mines to help ensure that the scrubber exhaust was not directed against the curtain line.

Use of deep-cutting techniques necessitates the use of greater curtain setback distances. Several research projects have been conducted to examine the effects of greater curtain setbacks on face area dust levels when using exhausting face ventilation. Mixed results have been reported. One laboratory study reported no change in dust levels in areas parallel to, and outby, the mouth of the exhausting curtain on the off-curtain side of the machine when the setback distance was increased from 30 to 40 ft [Colinet and Jankowski 1996]. The other laboratory study found significantly higher dust levels at the parallel position when increasing the setback from 30 to 40 ft [Goodman et al. 2006]. The machine-mounted water spray systems used for the two laboratory studies were substantially different and may have led to the discrepancies in findings. Another study, conducted underground, compared curtain setback distances of greater than 30 ft to distances less than 30 ft [Goodman 2000]. It found lower dust levels at the rear corners of the machine when using greater curtain setbacks. Based on these mixed results, it appears that the ideal curtain setback distance at mines using deep-cutting methods and exhausting face ventilation may be dependent on mine-specific water spray and ventilation parameters. In any case, to prevent the discharged scrubber airflow from striking the curtain line and blowing it against the rib, the curtain setback distance should be at least 30 ft. When the exhausting curtain is blown against the rib by the scrubber discharge, the airflow reaching the face is diminished.

Blowing Ventilation

From Table 1, dust levels in the shuttle car cabs when loading at the face during the blowing face ventilation cuts conducted at Mines A, D, E, and F averaged 1.96 mg/m^3 during the regular-cut depth and 2.32 mg/m^3 during the deep-cut depth. This difference was not statistically significant (85% CIs). In fact, 13 of the 18 blowing face cuts experienced no significant change when comparing deep-cut dust levels to regular-cut levels.

The curtain airflows presented in Table 1 for the blowing face ventilation cuts were measured after activation of the scrubber. This is an important distinction to make because both this field study and another laboratory study found that curtain airflow, as measured at the face-side mouth of the blowing curtain, can increase significantly upon activation of the scrubber [Taylor et al. 1997]. When the scrubber airflow and pre-activation blowing curtain airflow were similar, curtain airflows increased from 40% to 56% upon activation of the scrubber. At the blowing face ventilation mines surveyed for this study, the average ratio of airflow in the last open crosscut to the highest measured scrubber airflow ranged between 1.9 and 5.8. Since there was an abundance of airflow available in the last open crosscut, the additional air measured behind the curtain upon activation of the scrubber was likely a combination of re-circulated and fresh air. From a dust perspective, the curtain airflow reading after activation of the scrubber at these mines is important because it represents the actual airflow forced to the face during operating conditions.

As shown in Table 1, dust levels in the face area at Mine D increased as the depth of cut increased during the heading cut when using blowing face ventilation. This may be explained by a suspected less than ideal curtain-to-scrubber airflow ratio and may be due to the fact that the magnitude of scrubber airflow measured after this particular cut was the lowest observed value for blowing curtain cuts at 5,100 cubic ft per min (cfm). A laboratory study conducted by the USBM found that the ideal curtain-to-scrubber airflow ratio for controlling dust when using blowing face ventilation is 1.0, when curtain airflow is measured before activation of the scrubber [Jayaraman et al. 1992]. As mentioned previously, curtain airflows on the blowing faces surveyed for this study were measured after scrubber activation. The ratio measured at Mine D after completion of the cut, and with the scrubber running, was 1.82, the second highest value observed during the study. Assuming that the scrubber boosted airflow at Mine D was similar to what occurred during the Taylor study [1997], which had similar post activation scrubber (6,000 cfm) and curtain airflows (8,600 cfm), the estimated ending curtain-to-scrubber airflow ratio for the Mine D with the scrubber off is 1.30, which is a significant deviation from the ideal ratio of 1.0. The Jayaraman study also found that decreasing the scrubber airflow from 7,500 to 4,000 cfm was the most significant factor for increasing dust at all positions. It is interesting to note that scrubber airflows during this particular cut at Mine D dropped from 7,300 to 5,100 cfm, which was very close to laboratory conditions. Cut No. 18 at Mine E also experienced a substantial drop in scrubber airflow over the course of the cut but did not experience higher dust levels; however, it appears that the ending curtain-to-scrubber airflow ratio at Mine E was more favorable than at Mine D.

At Mine E, one of the four heading cuts using blowing face ventilation experienced significantly higher dust levels during the deep-cut depth. This was most likely caused by a change in shuttle car routes during the cut, which placed the cabs directly in the mining machine's return airflow for the last half of the cut. At Mine F, two of the eight heading cuts using blowing ventilation experienced significantly higher dust levels during the deep-cut depth; however, no determination could be made concerning the cause for this difference, as similar operating conditions were observed during all cuts. As shown in Appendix F, when all of the data collected at Mine F was combined for analysis, no significant (85% CIs) difference was found when comparing deep-cut dust levels to regular-cut levels. For the blowing ventilation cuts at Mine A, neither of the individual heading cuts demonstrated significant changes in dust during the deep-cut depth; however, when the data were combined, the dust levels were significantly (85% CIs) higher during the deep-cut depth. Higher deep-cut dust levels when driving headings were offset by lower deep-cut dust levels during the right crosscut development at Mine A, resulting in an overall change in dust levels that was not statistically significant (85% CIs) when comparing the regular-cut depth to the deep-cut depth.

A stepwise linear regression analysis, similar to the one conducted for the exhausting face data, was conducted on the blowing face ventilation data contained in Table 1. Again, calculations of the values for kurtosis (2.1) and skewness (1.3) associated with the dependent variable did not indicate a substantial deviation from an assumption of normality. The stepwise regression analysis found that only the identifier for Mine A was a significant independent variable (95% CIs), indicating that Mine A dust levels were significantly higher than those for the other three blowing face ventilation mines. As with the exhausting ventilation data set; collinearity existed in the airflow related independent variables. Additional regression analyses that involved separately comparing each independent variable to the dependent variable cannot rule out curtain and scrubber airflows as cause-and-effect variables, however, the mine identifier variable explained most of the variability in the dust levels.

For the blowing ventilation cuts shown in Table 3, the average miner operator exposure levels were 2.44 mg/m^3 during the regular-cut depth and 2.10 mg/m^3 during the deep-cut depth. This difference is not statistically significant (85% CIs). Mines A, D, and E properly positioned the miner operator at the mouth of the blowing curtain [Goodman and Listak 1999]. Mine F's approved ventilation plan precluded this positioning. The plan, which allows for 30-ft cuts, requires that the blowing curtain be within 10 ft of the face at the start of the cut and, after advancing the face 10 ft, requires its attachment to the last row of bolts. With these requirements, it was impossible for the miner operator to stand at the mouth of the curtain, resulting in higher dust exposure levels for the miner operator. During the last day of the survey at Mine F, the curtain setback distances exceeded plan, and the miner operator was able to position himself at the mouth of the blowing curtain. As a result, miner operator exposure levels dropped 3.94 mg/m^3 to an average level of 0.23 mg/m^3 during production. From a dust control perspective, the allowable curtain setback should be greater than the maximum cutting depth to ensure that miner operators can maintain their position at the mouth of the blowing curtain.

Several research projects have been conducted to examine the effects of greater curtain setbacks on face area dust levels when using blowing face ventilation. Two laboratory studies [Goodman 2000 and Jayaraman et al. 1992] found that increasing the curtain setback distance (20 to 40 ft in the Goodman study and 25 to 35 ft in the Jayaraman study) reduced dust rollback at the rear corners of the mining machine when using blowing ventilation. An underground blowing face study reported that dust levels at the right rear (operator's cab) of the mining machine initially increased when increasing the curtain setback distance from 25 to 35 ft but then decreased when increasing the setback from 35 to 50 ft [Volkwein et al. 1985]. Dust levels during the dustier slab cut were actually slightly lower when comparing the 50-ft setback to the 25-ft setback. The Volkwein study also found no degradation in methane dilution when increasing curtain setbacks from 25 to 50 ft. The results of these studies indicate that the greater curtain setbacks associated with deep-cutting methods at mines using blowing curtain were not detrimental to effective face ventilation.

No-curtain Ventilation

At mines using deep-cutting methods, greater curtain setback distances may result in cuts that do not require the installation of line curtain, such as the initial entry developments beyond the last open crosscut. Mines with smaller pillar dimensions will experience more of these cuts because the distance between the crosscuts is less. From Table 1, dust levels in the shuttle car cabs when loading at the face during the no-curtain ventilation cuts conducted at Mines B, D, and E averaged 0.46 mg/m^3 during the regular-cut depth and 0.15 mg/m^3 during the deep-cut depth. This difference is statistically significant (95% CIs). In fact, 9 of the 15 no-curtain cuts experienced significantly lower dust levels during the deep-cut depth when compared to the regular-cut depth. Lower shuttle car cab dust levels during the deep-cut portion of the no-curtain cuts can be explained by better scrubber inlet capture efficiencies and improved shuttle car positioning relative to the scrubber exhaust and intake airway.

As described earlier, area sampling was conducted upwind and downwind of the mining machine to determine dust generation during the regular- and deep-cut sequences. As seen in Appendix D, the mining machine at Mine D generated significantly (85% CIs) less dust during the deep-cut depth when compared to the regular-cut depth for cuts that did not use ventilation curtain. Assuming constant return ventilation airflow throughout each particular cut, lower dust generation indicates improved scrubber capture during the deep-cut depth. Improved capture occurs because the airflow reaching the face is essentially limited to what the scrubber is drawing after the miner extends beyond the influence of primary ventilation. In essence, a perfect

1.0 curtain-to-scrubber airflow ratio is created. In addition, primary ventilation airflow is usually perpendicular to the direction of cut during no-curtain cuts. The dustiest portions of the cuts occur when the miner is beginning the first sump and slab cuts and primary airflow sweeps across the cutting drum and under the boom of the miner. If the primary ventilation air velocities are high enough, scrubber inlet capture efficiencies decrease. Lower dust generation during the deep-cut depth partially explains lower shuttle car cab dust levels; however, shuttle car cab position is also a contributing factor.

As shown in Figure 3, when beginning the sump and slab lifts in no-curtain heading cuts, the scrubber exhaust is often located in or directed outby the last open crosscut, resulting in dust rollback into the shuttle car haulageway. This rollback can result in significant shuttle car operator exposures because the airflow in the haulageway is often low due to the placement of check and fly curtains in the entries. As shown in Figure 4, advancement of the cut results in the scrubber exhaust being directed into the last open crosscut, where the dusty air is quickly swept away from the workings, thereby reducing shuttle car operator exposure levels. Further face advancement places the shuttle car cab directly into the intake airway during the deep cut.

For the no-curtain cuts shown in Table 3, the average miner operator exposure levels were very low, 0.24 mg/m^3 during the regular-cut depth and 0.11 mg/m^3 during the deep-cut depth. This difference is not statistically significant (85% CIs). Due to the close proximity of no-curtain cuts to the last open crosscuts, the miner operators are able to position themselves in the intake airway, resulting in very low dust exposure levels.

Bolting Operations

When using deep-cut mining techniques, bolting operations are also subjected to longer cycle times. A buildup of material in the dust collection box may lower the suction pressure of the dry dust collection system, increasing operator exposure levels and bolting machine dust generation. One objective of this study was to determine if the bolting machine's dust collection efficiency decreased during a typical deep-cut bolting cycle. This was done by measuring the suction pressure of the bolter's dust collection system before and after each bolting cycle. Table 4 shows curtain airflows, face ventilation type, dust collector suction pressures, operator exposure levels, and bolter machine-generated dust levels during the regular- and deep-cut depths for 29 bolting cycles monitored for the study. Only bolting cycles conducted upwind of the miner were analyzed in Table 4 because miner generated dust, when compared to bolter generated dust, accounted for the majority of bolter operator exposures during downwind cycles. Mine C lists the bolter suction as nonapplicable (N/A) because the bolting machine used wet drilling techniques as opposed to a dry collection system. No attempt was made to analyze bolter machine dust generation during no-curtain cycles due to uncertainty related to determining a representative location for the return sampling package. Missing data for each mine are described in the Appendices.

Comparing only bolting cycles that had both left- and right-side data available, no statistically significant (85% CIs) difference was observed in the average dust exposure levels between the left-side operator at 0.45 mg/m^3 and the right-side operator at 0.48 mg/m^3. Therefore, all subsequent analyses group these data (left and right) together. As shown in Table 4, no degradation in bolter suction pressure was observed during individual bolting cycles. As a result, no statistically significant (85% CIs) difference was observed when comparing the average operator exposure level of 0.40 mg/m^3 during regular-cut depth to 0.58 mg/m^3 during deep-cut depth. Similar to the left- and right-side analysis, only operator exposure levels that had both regular- and deep-cut data available were compared. The only significant (95% CIs) difference occurred when comparing exposures for ventilated rooms to non-ventilated rooms. The average operator exposure level for ventilated room cycles was 0.40 mg/m^3, while the exposure for non-ventilated rooms was 0.96 mg/m^3. Ventilating the rooms with an average air quantity of 5,800 cfm reduced the average operator exposure level by approximately 0.56 mg/m^3, which underscores the importance of maintaining line curtain airflows during the bolting cycles. Similarly, no statistically significant (85% CIs) difference was observed in bolter dust generation when comparing regular- to deep-cut depth. Both were low at 0.12 mg/m^3 for regular-cut depth and 0.14 mg/m^3 for deep-cut depth.

The intra-mine analyses presented in the Appendices led to similar conclusions in that no significant (85% CIs) changes in bolting operation dust levels were found when comparing the regular- to the deep-cut portion of the cycle. A few general observations were made that may lead to improved practices at deep-cut mines. First, at Mine A, although only one bolting cycle was involved, dust levels appeared to have increased as the bolting cycle progressed when curtain ventilation was not measureable. This dust accumulation effect was not observed for ventilated cuts at Mine A. A similar upward trend in dust levels was observed at Mine F, where the blowing curtain was initially set and then not advanced with the progress of the bolting machine. As the distance between the bolter and the mouth of the curtain increased, less air may have reached the operators for dilution of drill-generated dust. These observations, while not statistically significant (85% CIs), reinforce prudent practices of maintaining measureable airflow to the bolting rooms, as well as periodically advancing the line curtain to assure that fresh air reaches the work areas.

Recent Developments Using Blocking Sprays

As shown in the Appendices, none of the mines surveyed for this study made use of blocking sprays to lower respirable dust exposures. When using exhausting face ventilation and a flooded-bed scrubber, a laboratory study demonstrated that blocking sprays can significantly lower dust levels on the off-curtain side of the mining machine during the slab cut [Organiscak and Beck, forthcoming]. For this lab study, two hollow cone sprays were located on each side of the mining machine body approximately 2 ft outby the scrubber inlet and 2 ft above ground level. The two sprays were 3-in apart on a vertical axis and were angled 15 degrees away from the mining machine body and toward the face. For this particular study, hollow cone sprays were found to be better than flat fan sprays for controlling airborne dust. When using blowing face ventilation and a flooded-bed scrubber, a study demonstrated that blocking sprays improved suppression of the dust cloud underneath the boom of the miner in the laboratory and lowered dust levels at the rear corners of the machine and at the operator's position during field testing [Goodman 2000]. The laboratory and field blocking spray systems designed for the Goodman study used vertically oriented flat fan sprays instead of hollow cone sprays, which were not tested. Results of these studies indicate that further dust reductions may be achieved on continuous mining sections through the implementation of blocking spray technology on mining machines equipped with flooded-bed scrubbers.

Conclusions

All of the six mines surveyed for this study were able to successfully implement deep-cut mining practices without significantly increasing the dust exposures of face workers during cutting and bolting cycles. Researchers observed the implementation of mining practices that have previously been shown to benefit dust control during deep-cut depth, and, in some cases, they were able to demonstrate the importance of these measures. While most known flooded-bed scrubber and deep-cut dust control practices appear to be widely implemented by the mining industry, none of the operations surveyed for this study used blocking sprays on the mining machines to improve dust control. Studies have demonstrated the benefits of using blocking sprays to control dust on both exhausting [Organiscak and Beck, forthcoming] and blowing [Goodman 2000] faces, indicating that implementation of this technology could further improve industry-wide respirable dust exposure levels.

Field data indicated that the most important factor for controlling dust on the exhausting faces was scrubber airflow volume. This fact, combined with the observation that 22% of the deep cuts surveyed for this study experienced a drop in scrubber airflow in the range of 20% to 35%, suggests that the scrubber screens should be cleaned more frequently than the common manufacturer recommendation of twice a shift. Screens should be tapped and back-flushed before each deep cut. All of the mines surveyed for this study used 20-mesh scrubber screens. This study also verified that, when using exhausting face ventilation, the shuttle car cabs should be located on the off-curtain side of the entry to help ensure that the shuttle car operator exposure levels do not increase during the deep cut. In general, all the exhausting face ventilation mines employed proper miner operator positioning at a location parallel to or outby the face-side mouth of the curtain [Colinet and Jankowski 1996]. Research is inconclusive on the proper curtain setback distance for deep cuts with exhausting face ventilation [Colinet and Jankowski 1996; Goodman et al. 2006; Goodman 2000], and the ideal setback distance is likely dependent on mine-specific water spray and ventilation conditions. However, the setback distance should be great enough to prevent the scrubber exhaust from blowing the line curtain against the rib. This distance is generally greater than 30 ft.

When using blowing face ventilation, the importance of maintaining the proper curtain-to-scrubber airflow ratio of 1.0 was apparent during one cut when this ratio deteriorated (increased) and dust levels sharply increased. The ideal airflow ratio of 1.0 was determined in the laboratory, where curtain airflow was measured before activation of the scrubber [Jayaraman et al. 1992]. Laboratory [Taylor et al. 1997] and field testing observed 40% to 54% increases in curtain airflow upon activation of the scrubber when the scrubber capacity and pre-activation curtain airflows were similar. Therefore, mine ventilation plans should stipulate that curtain airflow be tested before activation of the scrubber to avoid erroneously inflating the curtain-to-scrubber airflow ratio. Mine practices should also ensure sufficient curtain setback distance to allow miner operators to maintain their position at the face-side mouth of the curtain for the entire cut. In order to achieve this objective, a curtain setback distance that is greater than the allowed deep-cut depth would have to be permitted. Several studies have demonstrated no degradation in face ventilation effectiveness when using curtain setbacks of up to 50 ft [Goodman 2000; Jayaraman et al. 1992; Volkwein et al. 1985]. The curtain setback requirement contained in the ventilation plan at one mine surveyed for this study prevented the miner operator from standing behind the blowing curtain, where dust levels were over 90% lower than in the entry. Miner operator exposure levels at the lesser curtain setback mine were higher than at the other mines, where the operators spent the majority of their time at the mouth of the curtain.

At mines using deep-cutting methods, the use of greater curtain setback distances results in some cuts that do not require the installation of curtain. For these cuts, dust levels were lower during deep-cut depth when compared to regular-cut depth due to improved worker positioning and better scrubber inlet capture efficiencies as the cuts progressed. However, effective ventilation of no-curtain cuts is entirely dependent upon proper scrubber functioning.

When adequately ventilated, dust exposures on the bolting faces and bolting machine dust generation did not appear to be affected by the longer bolting cycles associated with deep-cut mining practices at the mines surveyed for this study. However, although not statistically significant (85% CIs), some data did indicate the potential for higher exposures during longer bolting cycles if either curtain airflow was not measureable or the curtain was not advanced as the bolting cycle progressed. In addition, bolter operator exposure levels during bolting cycles that were ventilated with an average curtain airflow level of 5,800 cfm were 0.56 mg/m^3 lower than exposures when no face airflow was detected.

References

CFR. Code of Federal Regulations. Washington, DC: U.S. Government Printing Office, Office of the Federal Register.

Colinet JF, Jankowski RA [1996]. Dust control considerations for deep-cut faces when using exhaust ventilation and a flooded-bed scrubber. Littleton, CO: Society for Mining, Metallurgy, and Exploration, Inc., SME Transactions *302*:104–111.

Colinet JF, Rider JP, Listak JM, Chekan GJ [2010]. A summary of USBM/NIOSH respirable dust control research for coal mining. In: Brune J, ed. Extracting the science: a century of mining research. Littleton, CO: Society for Mining, Metallurgy, and Exploration, pp. 432–441.

EIA [2008]. Annual coal report 2008. Washington, DC: U.S. Department of Energy, Energy Information Administration, Report No. DOE/EIA 0584.

Goodman GVR [2000]. Using water sprays to improve performance of a flooded-bed dust scrubber. Appl Occup Environ Hyg *15*(7):550–560.

Goodman GVR, Listak JM [1999]. Variation in dust levels with continuous miner position. Min Eng *51*(2):53–59.

Goodman GVR, Beck TW, Pollock DE [2006]. The effects of water spray placement for controlling respirable dust and face methane concentrations. J Mine Vent Soc S Afr *April/June*: 56–63.

Jayaraman NI, Jankowski RA, Whitehead KL [1992]. Optimizing continuous miner scrubbers for dust control in high coal seams. In: Proceedings of New Technology in Mine Health and Safety, SME Annual Meeting. Littleton, CO: Society for Mining, Metallurgy, and Exploration, pp. 193–205.

McClelland JJ, Colinet JF, Bringhurst B[1992]. Performance evaluation of irrigated filters. Littleton, CO: Society for Mining, Metallurgy, and Exploration, Inc., SME Transactions *290*:1828-1831.

MSHA [2010]. Program Evaluation and Information Resources, Standardized Information System. Arlington, VA: U.S. Department of Labor, Mine Safety and Health Administration.

Niewiadomski G [2010]. Producing MMU's by method of mining and mines by type. Washington, DC: Mine Safety and Health Administration, Coal Mine Safety and Health, Health Division, telephone conversation with Colinet.

Organiscak JA, Beck TW [forthcoming]. Continuous miner spray considerations for optimizing scrubber performance in exhaust ventilation systems. Paper presented at the SME Annual Meeting, Phoenix, AZ, February 28–March 3, 2010, Preprint 10–204.

Taylor DT, Rider JP, Thimons ED [1997]. Impact of unbalanced intake and scrubber flows on face methane concentrations. In: Ramani RV, ed. Proceedings of the 6th International Mine Ventilation Congress. Chapter 27. Littleton, CO: Society for Mining, Metallurgy, and Exploration, pp. 169–172.

USBM [1970]. Proceedings of the Symposium on Respirable Coal Mine Dust. Washington, DC: Department of the Interior, U.S. Bureau of Mines, USBM Information Circular 8558, p. 253.

Volkwein JC, Thimons ED, Halfinger G [1985]. Extended advance of continuous miner successfully ventilated with a scrubber in a blowing section. In: Proceedings of the 2nd U.S. Mine Ventilation Symposium. New York: Taylor and Francis, pp. 741–745.

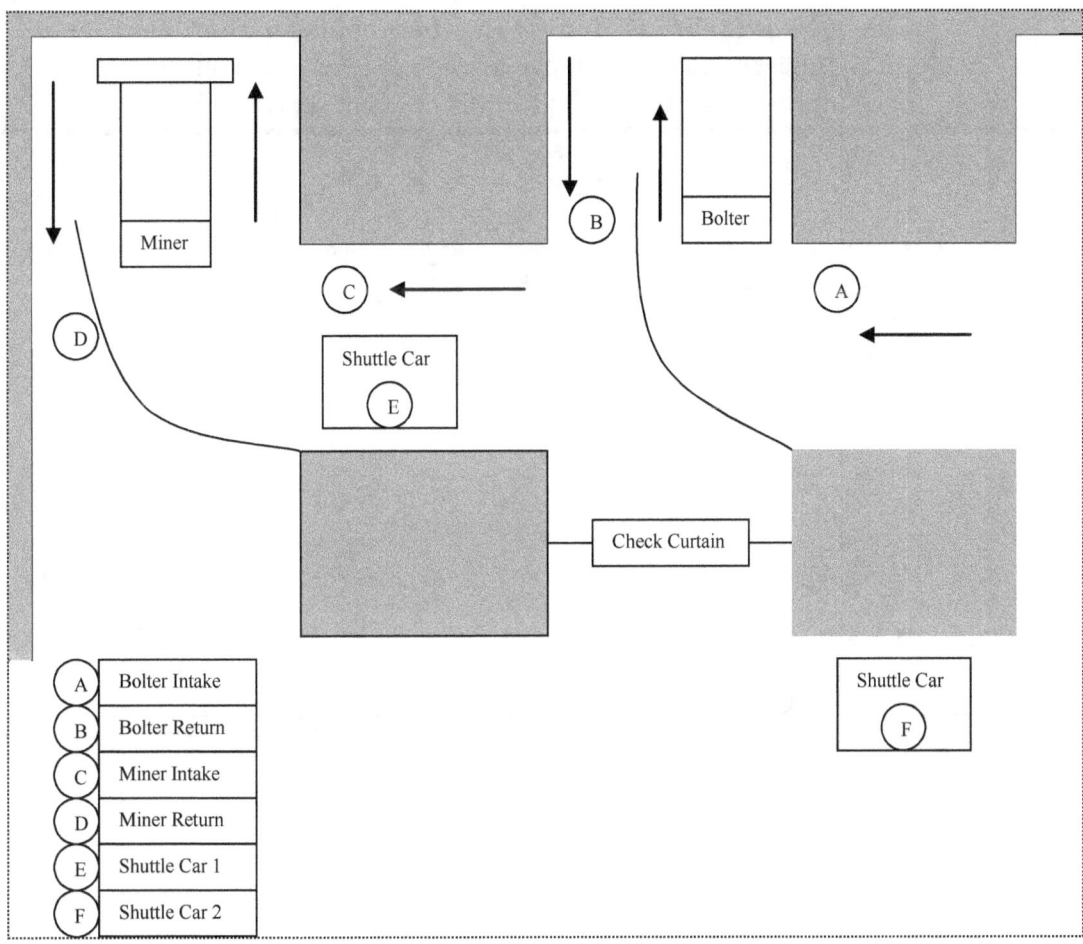

Figure 1. Area sampling locations for a typical section layout. Note: PDMs Are not shown.

Table 1. Shuttle car cab dust levels when loading at the face
Note: Significance column compares deep-cut levels to regular-cut levels by comparing the 85% CIs of the averages.

Cut No.	Face Ventilation	Orientation	Mine	Curtain Airflow (cfm)	Starting Scrubber Airflow (cfm)	Ending Scrubber Airflow (cfm)	Starting Airflow Ratio	Ending Airflow Ratio	Regular-Cut Dust Level (mg/m^3)	Deep-Cut Dust Level (mg/m^3)	Statistically Significant
1	Exhausting	Heading	A	10,100	11,100	10,000	0.91	1.01	0.16	0.28	No
2	Exhausting	Heading	A	13,500	11,700	11,700	1.15	1.15	0.38	1.04	No
3	Exhausting	Heading	A	10,900	11,700	11,100	0.93	0.98	0.06	0.35	No
4	Exhausting	Heading	B	8,100	7,200	7,200	1.13	1.13	0.52	0.81	No
5	Exhausting	Heading	B	5,100	7,400	4,800	0.69	1.06	0.78	0.39	No
6	Exhausting	Heading	B	9,000	7,600	6,900	1.18	1.30	0.40	0.26	No
7	Exhausting	Heading	C	21,500	13,300	12,700	1.62	1.69	0.05	0.37	No
8	Exhausting	Heading	C	16,700	13,700	12,400	1.22	1.35	0.07	0.16	No
9	Exhausting	Heading	C	18,300	14,000	13,200	1.31	1.39	0.04	0.24	Yes, Higher
10	Exhausting	Heading	C	17,300	14,000	12,700	1.24	1.36	0.07	0.25	Yes, Higher
11	Exhausting	Heading	C	17,200	14,200	13,200	1.21	1.30	0.08	0.11	Yes, Higher
12	Exhausting	Right Crosscut	C	16,600	0	12,700	N/A	1.31	0.03	0.46	Yes, Higher
13	Exhausting	Right Crosscut	C	17,100	13,200	12,900	1.30	1.33	0.15	0.11	No
14	Exhausting	Right Crosscut	C	14,300	13,500	12,900	1.06	1.11	0.07	0.12	No
15	Blowing	Heading	A	18,000	11,100	11,100	1.62	1.62	4.07	5.07	No
16	Blowing	Heading	A	13,000	10,700	10,700	1.21	1.21	4.38	7.34	No
17	Blowing	Heading	D	9,300	7,300	5,100	1.27	1.82	0.75	2.73	Yes, Higher
18	Blowing	Heading	E	6,700	8,500	5,500	0.79	1.22	1.48	1.08	No
19	Blowing	Heading	E	6,500	8,500	7,800	0.76	0.83	1.07	1.62	No
20	Blowing	Heading	E	Minimal	7,800	7,500	N/A	N/A	0.18	0.82	Yes, Higher
21	Blowing	Heading	E	Minimal	7,500	7,500	N/A	N/A	0.31	0.18	No
22	Blowing	Heading	F	6,500	8,500	8,200	0.76	0.79	0.10	0.30	Yes, Higher
23	Blowing	Heading	F	6,400	8,800	8,200	0.73	0.78	0.93	2.21	Yes, Higher
24	Blowing	Heading	F	7,100	9,100	8,500	0.78	0.84	2.52	3.00	No
25	Blowing	Heading	F	6,500	8,800	8,800	0.74	0.74	2.20	2.08	No
26	Blowing	Heading	F	7,100	9,100	8,500	0.78	0.84	2.61	2.08	No
27	Blowing	Heading	F	6,400	8,800	8,800	0.73	0.73	2.29	3.33	No
28	Blowing	Heading	F	6,800	8,700	8,300	0.78	0.82	2.66	2.27	No
29	Blowing	Heading	F	6,400	9,100	8,200	0.70	0.78	1.78	1.86	No
30	Blowing	Left Crosscut	E	5,400	8,800	8,300	0.61	0.65	1.49	1.10	No
31	Blowing	Right Crosscut	A	14,000	10,700	9,500	1.31	1.47	5.26	3.55	Yes, Lower
32	Blowing	Right Crosscut	E	17,800	8,500	8,500	2.09	2.09	1.24	1.13	No
33	No Curtain	Heading	B	0	7,000	6,900	0.00	0.00	0.06	0.05	No
34	No Curtain	Heading	B	0	7,700	6,100	0.00	0.00	0.39	0.63	No
35	No Curtain	Heading	D	0	7,000	4,600	0.00	0.00	1.61	0.24	Yes, Lower
36	No Curtain	Heading	D	0	7,100	4,800	0.00	0.00	0.63	0.06	Yes, Lower
37	No Curtain	Heading	D	0	7,200	6,000	0.00	0.00	0.17	0.03	Yes, Lower
38	No Curtain	Heading	D	0	6,900	5,700	0.00	0.00	0.57	0.11	Yes, Lower
39	No Curtain	Heading	D	0	7,100	5,400	0.00	0.00	0.29	0.03	Yes, Lower
40	No Curtain	Heading	D	0	7,800	6,300	0.00	0.00	0.31	0.05	Yes, Lower
41	No Curtain	Heading	E	0	7,800	5,500	0.00	0.00	0.43	0.02	Yes, Lower
42	No Curtain	Heading	E	0	6,800	7,100	0.00	0.00	0.32	0.10	No
43	No Curtain	Heading	E	0	7,100	7,100	0.00	0.00	0.16	0.03	Yes, Lower
44	No Curtain	Heading	E	0	7,800	7,800	0.00	0.00	0.20	0.16	No
45	No Curtain	Left Crosscut	B	0	7,900	7,500	0.00	0.00	0.25	0.12	No
46	No Curtain	Right Crosscut	D	0	7,200	4,700	0.00	0.00	0.71	0.29	Yes, Lower
47	No Curtain	Right Crosscut	D	0	7,100	5,100	0.00	0.00	0.76	0.31	No

Table 2. Time weighted average levels for the shuttle car, miner, and bolter operators

Mine	Day	Shuttle Car Cab 1 mg/m³	Shuttle Car Cab 2 mg/m³	Shuttle Car Cab 3 mg/m³	Miner Operator mg/m³	Left-Side Bolter Operator mg/m³	Right-Side Bolter Operator mg/m³
A	1	0.45	0.53	-	0.76	0.99	1.06
A	2	1.16	1.23	-	1.76	-	0.91
B	1	0.17	0.25	-	0.34	0.30	0.82
B	2	0.35	0.32	-	0.96	0.52	0.45
C	1	0.15	0.16	-	0.36/0.57	0.55	-
C	2	0.39	0.21	-	0.34/0.39	0.16	0.14
C	3	0.13	0.15	-	0.37/0.49	0.17	0.11
D	1	0.20	0.53	-	0.08	0.11	0.12
D	2	0.11	0.35	-	0.02	0.05	0.07
D	3	0.19	0.43	-	-	0.20	0.22
E	1	0.39	0.38	0.42	-	0.52	0.80
E	2	0.22	0.17	0.18	-	0.83	0.82
E	3	0.10	0.16	0.11	-	-	-
F	1	0.21	0.48	-	-	-	-
F	2	0.47	0.61	-	1.71	0.61	0.61
F	3	0.27	0.50	-	0.81	0.95	1.02

Table 3. Miner operator exposure levels and miner-generated dust levels during regular- and deep-cuts depths

Cut No.	Face Ventilation	Orientation	Mine	Regular-Cut Miner Operator mg/m³	Deep-Cut Miner Operator mg/m³	Regular-Cut Miner Generated mg/m³	Deep-Cut Miner Generated mg/m³
1	Exhausting	Heading	A	0.39	0.91	2.77	1.69
2	Exhausting	Heading	A	1.57	1.76	1.53	2.16
3	Exhausting	Heading	A	0.43	1.40	1.04	2.37
4	Exhausting	Heading	B	0.25	0.35	0.61	0.59
5	Exhausting	Heading	B	1.11	0.90	0.73	1.31
6	Exhausting	Heading	B	0.58	0.70	5.85	2.19
7	Exhausting	Heading	C	0.30	0.57	1.32	2.39
8	Exhausting	Heading	C	0.44	0.49	1.33	1.01
9	Exhausting	Heading	C	0.17	0.37	1.54	1.43
10	Exhausting	Heading	C	0.33	0.89	3.34	1.99
11	Exhausting	Heading	C	0.39	0.45	1.79	1.71
12	Exhausting	Right Crosscut	C	0.79	1.32	-	-
13	Exhausting	Right Crosscut	C	0.84	0.47	-	-
14	Exhausting	Right Crosscut	C	0.08	0.97	1.32	4.62
15	Blowing	Heading	A	0.67	1.04	1.93	1.60
16	Blowing	Heading	A	1.45	1.41	2.54	4.03
17	Blowing	Heading	D	0.03	0.03	1.73	1.67
18	Blowing	Heading	E	-	-	0.08	0.00
19	Blowing	Heading	E	-	-	0.47	0.83
20	Blowing	Heading	E	-	-	0.62	0.81
21	Blowing	Heading	E	-	-	0.99	0.55
22	Blowing	Heading	F	-	-	1.05	1.15
23	Blowing	Heading	F	-	-	0.99	1.07
24	Blowing	Heading	F	-	-	0.76	0.78
25	Blowing	Heading	F	3.70	4.18	0.21	0.17
26	Blowing	Heading	F	4.09	3.22	0.53	0.29
27	Blowing	Heading	F	4.95	4.85	1.09	0.84
28	Blowing	Heading	F	0.26	0.31	1.00	0.91
29	Blowing	Heading	F	0.14	0.19	0.58	0.29
30	Blowing	Left Crosscut	E	-	-	0.61	0.48
31	Blowing	Right Crosscut	A	6.64	3.70	2.95	1.75
32	Blowing	Right Crosscut	E	-	-	1.35	1.04
33	No Curtain	Heading	B	0.20	0.20	1.82	1.48
34	No Curtain	Heading	B	1.14	0.21	16.89	1.78
35	No Curtain	Heading	D	0.03	0.04	3.97	1.05
36	No Curtain	Heading	D	0.09	0.09	4.56	0.60
37	No Curtain	Heading	D	0.05	0.05	1.54	0.68
38	No Curtain	Heading	D	0.16	0.09	2.66	1.33
39	No Curtain	Heading	D	0.06	0.10	5.08	1.46
40	No Curtain	Heading	D	0.03	0.03	2.62	0.52
41	No Curtain	Heading	E	-	-	1.28	0.47
42	No Curtain	Heading	E	-	-	0.85	0.28
43	No Curtain	Heading	E	-	-	2.75	1.00
44	No Curtain	Heading	E	-	-	1.14	0.43
45	No Curtain	Left Crosscut	B	0.69	0.34	1.18	1.76
46	No Curtain	Right Crosscut	D	0.06	0.05	0.55	0.20
47	No Curtain	Right Crosscut	D	0.12	0.03	1.50	0.69

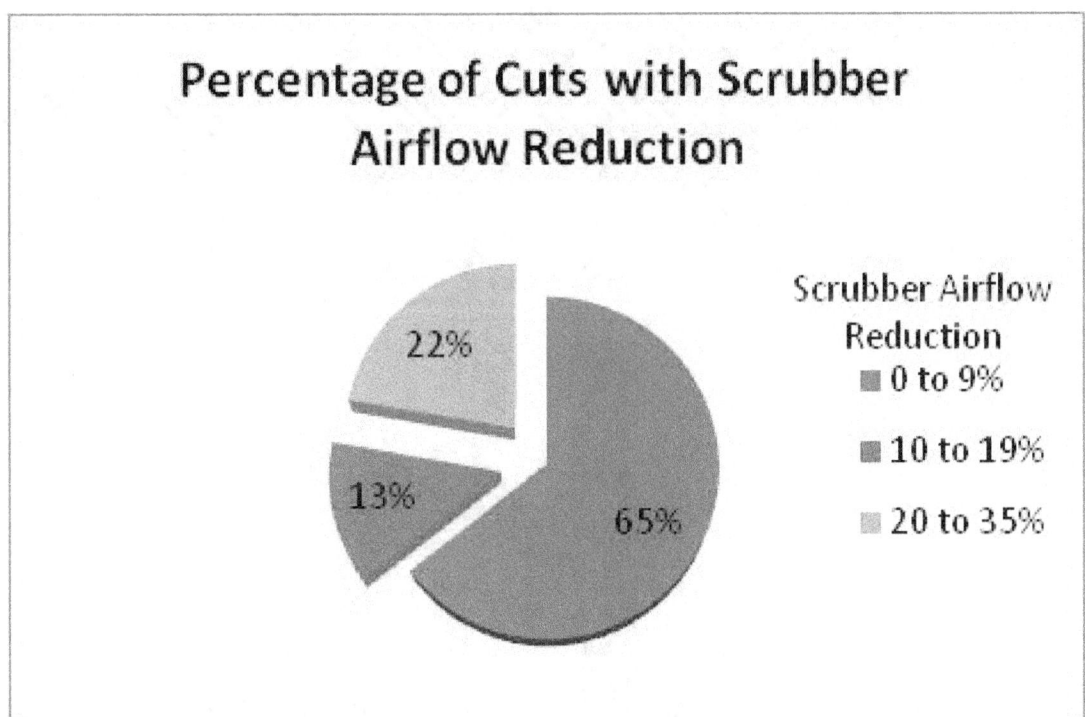

Figure 2. Pie chart depicting the magnitude of scrubber airflow reduction during the cut as a percentage of total cuts measured.

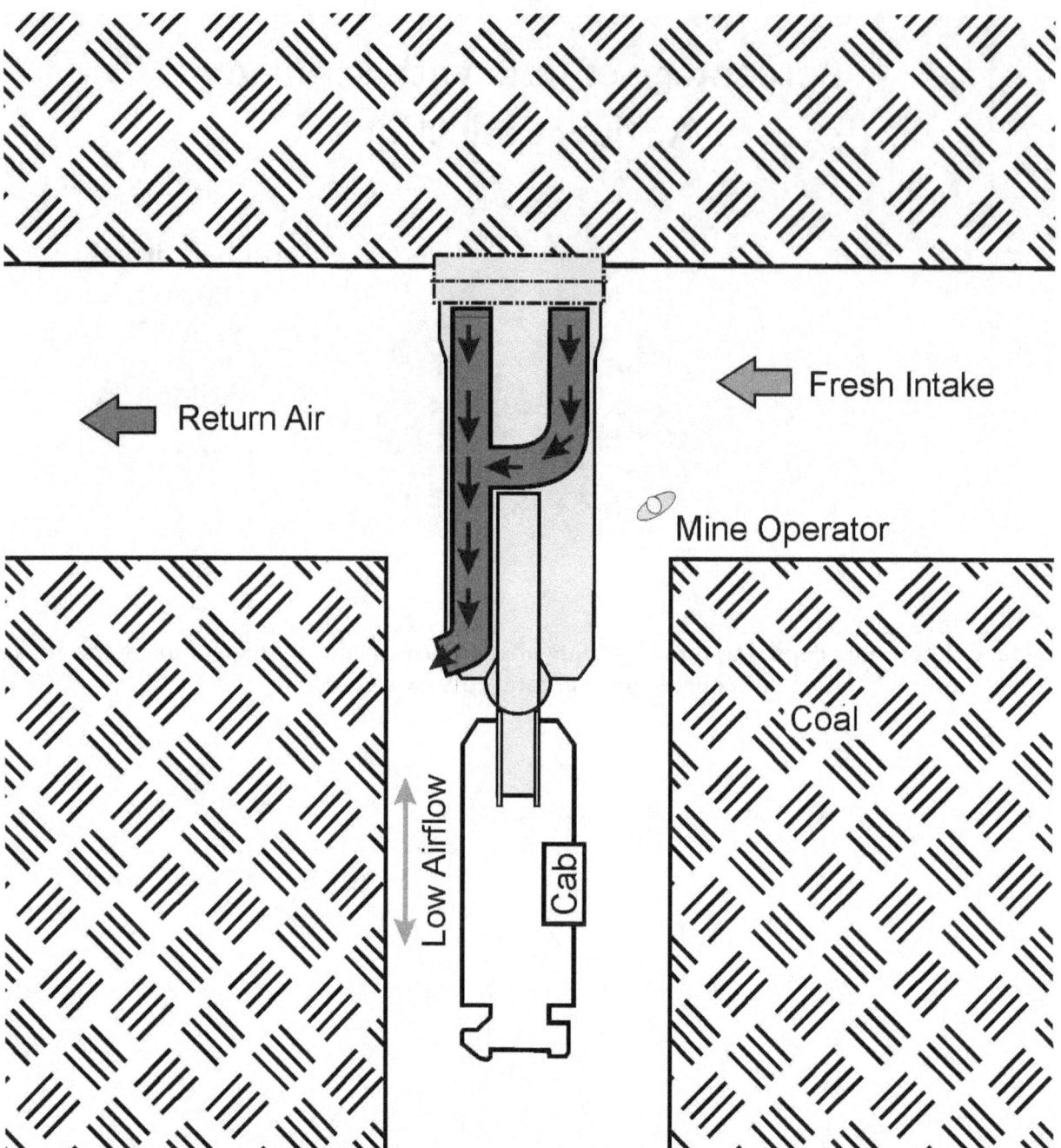

Figure 3. Mine operator and shuttle car cab position at the start of the no-curtain cut for the initial heading development beyond the last open crosscut.

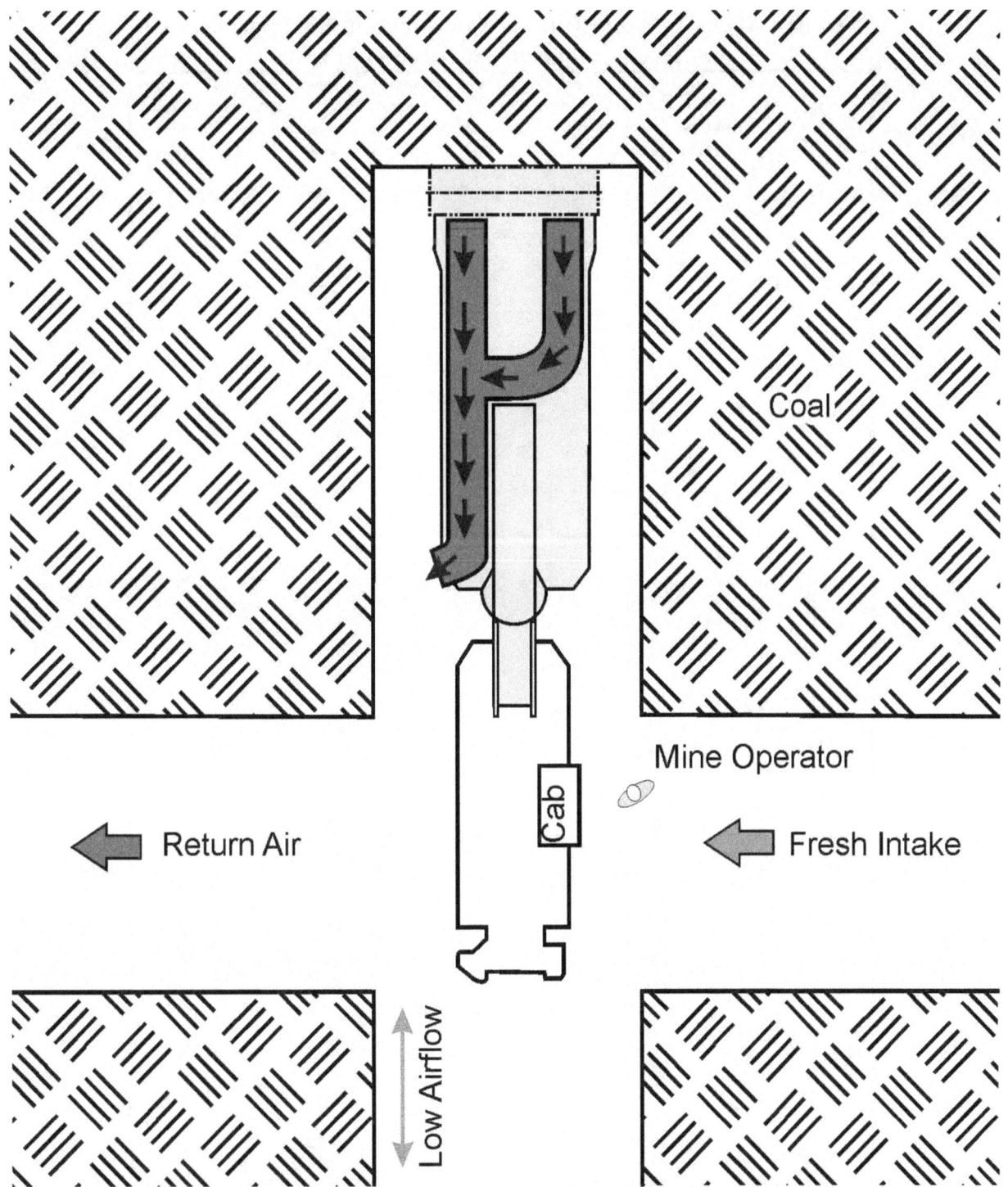

Figure 4. Mine operator and shuttle car cab position at the end of the no-curtain cut for the initial heading development beyond the last open crosscut.

Table 4. Bolter operator exposure levels and machine-generated dust levels at regular- and deep-cut depths

Mine	Room No.	Depth	Curtain Airflow (cfm)	Face Ventilation	Bolter Suction Left/Right (in Hg)	Left-Side Operator Dust Levels (mg/m³)	Right-Side Operator Dust Levels (mg/m³)	Upwind Bolter Dust Levels (mg/m³)	Downwind Bolter Dust Levels (mg/m³)	Bolter Generated Dust (mg/m³)
B	1	Regular	1,800	Exhausting	15/13	0.00	0.29	0.08	0.27	0.19
B	1	Deep	1,800	Exhausting	15/13	0.17	0.19	0.05	0.07	0.02
B	2	Regular	4,700	Exhausting	15/13	0.73	0.35	0.07	0.17	0.10
B	2	Deep	4,700	Exhausting	15/13	-	0.59	0.02	0.13	0.11
C	3	Regular	2,600	Exhausting	N/A	0.55	-	0.42	-	-
C	3	Deep	2,600	Exhausting	N/A	0.57	-	0.19	-	-
C	4	Regular	3,800	Exhausting	N/A	0.26	-	0.08	-	-
C	4	Deep	3,800	Exhausting	N/A	0.34	-	0.20	-	-
C	5	Regular	5,000	Exhausting	N/A	0.15	0.17	0.04	-	-
C	5	Deep	5,000	Exhausting	N/A	0.18	0.15	0.06	-	-
C	6	Regular	4,600	Exhausting	N/A	0.16	0.16	0.26	-	-
C	6	Deep	4,600	Exhausting	N/A	0.15	0.16	0.16	-	-
C	7	Regular	3,300	Exhausting	N/A	0.14	0.14	0.17	-	-
C	7	Deep	3,300	Exhausting	N/A	0.13	0.30	0.13	-	-
C	8	Regular	14,200	Exhausting	N/A	0.13	0.13	0.12	0.21	0.09
C	8	Deep	14,200	Exhausting	N/A	0.15	0.13	0.21	0.20	0.00
C	9	Regular	6,400	Exhausting	N/A	0.26	0.13	0.08	0.13	0.05
C	9	Deep	6,400	Exhausting	N/A	0.13	-	0.10	0.11	0.02
A	10	Regular	3,400	Blowing	12/12	-	1.39	0.00	0.04	0.04
A	10	Deep	3,400	Blowing	12/12	-	1.70	0.00	0.06	0.06
A	11	Regular	5,200	Blowing	12/12	-	0.91	0.00	0.13	0.13
A	11	Deep	5,200	Blowing	12/12	-	0.74	0.00	0.11	0.11
D	12	Regular	5,200	Blowing	17/17	0.00	0.00	0.11	0.09	0.00
D	12	Deep	5,200	Blowing	17/17	0.24	0.24	0.15	0.12	0.00
D	13	Regular	3,700	Blowing	17/17	0.17	0.16	0.02	0.11	0.09
D	13	Deep	3,700	Blowing	17/17	0.00	0.00	0.02	0.20	0.18
D	14	Regular	9,000	Blowing	18/18	0.00	0.00	0.24	0.12	0.00
D	14	Deep	9,000	Blowing	18/18	0.00	0.00	0.37	0.08	0.00
D	15	Regular	8,600	Blowing	17/17	0.00	0.00	0.02	0.09	0.07
D	15	Deep	8,600	Blowing	17/17	0.00	0.00	0.03	0.16	0.13
D	16	Regular	2,600	Blowing	18/17	0.00	0.11	0.03	0.12	0.09
D	16	Deep	2,600	Blowing	18/17	0.00	0.00	0.05	0.13	0.08
D	17	Regular	7,600	Blowing	17/17	0.18	0.00	0.08	0.11	0.03
D	17	Deep	7,600	Blowing	17/17	0.00	0.36	0.12	0.10	0.00
D	18	Regular	2,000	Blowing	16/17	0.00	0.15	0.02	0.15	0.13
D	18	Deep	2,000	Blowing	16/17	0.35	0.00	0.07	0.22	0.15
D	19	Regular	7,000	Blowing	17/18	0.00	0.00	0.02	0.08	0.06
D	19	Deep	7,000	Blowing	17/18	0.00	0.00	0.05	0.09	0.05
D	20	Regular	7,000	Blowing	16/17	0.22	0.00	0.16	0.16	0.00
D	20	Deep	7,000	Blowing	16/17	0.00	0.40	0.25	0.15	0.00
F	21	Regular	7,400	Blowing	15/14	-	-	0.03	0.21	0.18
F	21	Deep	7,400	Blowing	15/14	-	-	0.02	0.19	0.17
F	22	Regular	5,500	Blowing	13/12	2.20	0.53	0.02	-	-
F	22	Deep	5,500	Blowing	13/12	0.30	0.35	0.01	-	-
F	23	Regular	10,300	Blowing	13/14	0.85	0.51	0.01	0.19	0.18
F	23	Deep	10,300	Blowing	13/14	2.16	2.02	0.00	0.47	0.47
F	24	Regular	4,900	Blowing	14/13	0.63	1.47	0.08	0.50	0.42
F	24	Deep	4,900	Blowing	14/13	1.99	5.49	0.02	1.01	0.98
F	25	Regular	9,100	Blowing	14/13	0.43	0.30	0.02	0.39	0.37
F	25	Deep	9,100	Blowing	14/13	0.77	0.25	0.08	0.30	0.22
A	26	Regular	None	None	12/12	0.65	0.96	-	-	-
A	26	Deep	None	None	12/12	2.65	1.61	-	-	-
E	27	Regular	None	None	10/11	1.06	1.01	-	-	-
E	27	Deep	None	None	10/11	0.89	0.60	-	-	-
E	28	Regular	None	None	11/8	0.00	0.40	-	-	-
E	28	Deep	None	None	11/8	0.28	0.88	-	-	-
F	29	Regular	None	None	12/12	1.70	1.13	-	-	-
F	29	Deep	None	None	12/12	0.75	0.75	-	-	-

Appendix A: MINE A CASE STUDY

Mine-specific Information

Mine A used a Joy 12CM12 continuous miner. The miner spray configuration is shown in Figure A-1. A total of 38 dust suppression sprays were operated at a pressure of not less than 70 pounds per square inch (psi) as required by the mine ventilation plan. The sprays used in spray blocks 1, 3, 4, 6, and 7 were Model 6508 and those in spray blocks 2, 5, 8, 9, and 10 were Model 6512. The ventilation plan required the scrubber capacity to be a minimum of 9,000 cfm using a 20-layer (12-in x 26-in) scrubber filter. The duct work spray was a full-cone spray with a 106° angle and an orifice size of 0.234-in. It operated at 6.4 gal per min (gpm) at 45 psi.

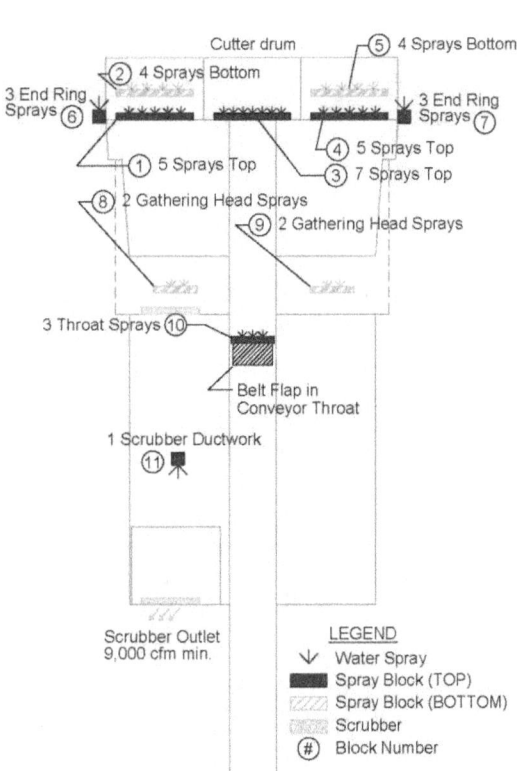

Figure A-1. Mine A continuous miner spray configuration.

The cut sequence for Mine A was as follows:

(1) 20-ft sump cut, right side
(2) 20-ft slab cut, left side
(3) 20-ft sump cut, right side
(4) 20-ft slab cut, left side

Figure A-2 shows a plan view of the mining section of Mine A at the time of this study. When cutting on the right side of the section (Entries No. 5, 6, and 7), the mine used blowing curtain ventilation. The mine used exhausting curtain when cutting on the left side (Entries No. 1, 2, 3). Entry No. 4 could be ventilated with either technique. The bolting operations also used blowing and exhausting curtain ventilation depending on their location, similar to the continuous miner. Mine A was operating sweep air ventilation, with Entry No. 7 serving as the intake and Entry No. 1 serving as the return. Entries No. 2 through No. 6 were neutral. The belt was in Entry No. 4. The main intake airflow averaged 38,700 cfm. Mining height was approximately 10 ft with entry width averaging 20 ft. There were two partings in the coal seam consisting of a 6-in band of bottom slate and an 18-in band of slate approximately 16-in below the roof.

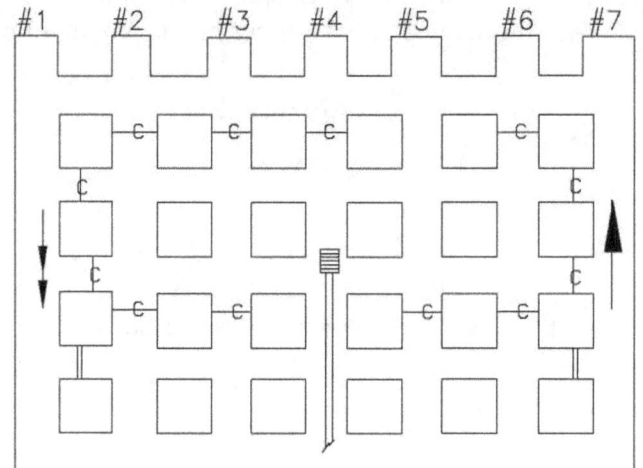

Figure A-2. Plan view of the mine A operating section.

The dust survey for Mine A was conducted over 3 consecutive days from February 3, 2009, to February 5, 2009. The third day of the study did not yield any data because the continuous miner was down due to low water-spray pressure.

Shuttle Car Data Analysis

Tables A-1, A-2, and A-3 show the shuttle car data collected during the study. On Day 1 of the study, the mine used three shuttle cars; however, only two were sampled. Therefore, the loading periods for the third shuttle car do not contain dust concentration data as indicated in the tables. On Day 2 of the study, the mine only used two shuttle cars; therefore, dust concentrations for all of the loading periods are presented. Dust exposures were analyzed for the following unique deep-cut conditions:

1. Continuous miner driving a heading with exhausting face ventilation
2. Continuous miner driving a heading with blowing ventilation
3. Continuous miner turning a right crosscut with blowing ventilation

Figure A-3 is a plot of shuttle car dust exposures for three cuts under condition 1—continuous miner driving a heading with exhausting face ventilation. The x-axis corresponds to the sequential load number during the cut. There were 18 to 19 loads per 40-ft cut; therefore, load 9 represents the transition from regular- to deep-cut depth. When driving a heading using exhausting face ventilation, shuttle car dust levels were very low during both the regular- and the deep-cut sequence, averaging 0.19 mg/m^3 during the regular-cut sequence and 0.56 mg/m^3 during the deep-cut sequence. In fact, the difference between these averages was not statistically significant (85% CIs), indicating that using deep cutting methods did not impact shuttle car dust levels under condition 1. A spike in the shuttle car cab's dust level was observed during the 15th load in Entry No. 2. The reason for this spike is unknown.

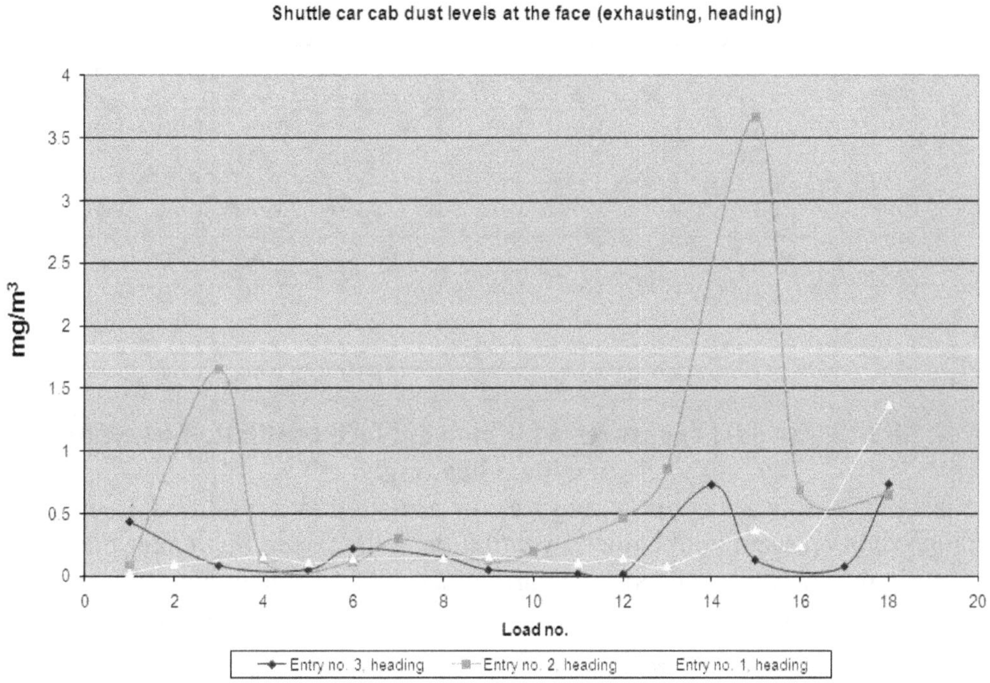

Shuttle car cab dust levels at the face (exhausting, heading)

Figure A-3. Shuttle car dust exposures with exhausting face ventilation when the miner was driving a heading.

Figure A-4 is a plot of shuttle car dust exposures for two cuts under condition 2—continuous miner driving a heading with blowing ventilation. Data collected during development of Entry No. 5, as shown in Table A-2, are not included in the plot because it was only 25 ft deep, resulting in an insufficient number of deep-cut data points. When driving a heading using blowing ventilation, dust exposures at the shuttle cars increased during loading in the deep-cut sequence. Dust levels averaged 4.08 mg/m^3 when loading during the regular-cut depth and 6.38 mg/m^3 during the deep-cut depth. The 85% CIs for these averages did not overlap, indicating a statistically significant difference.

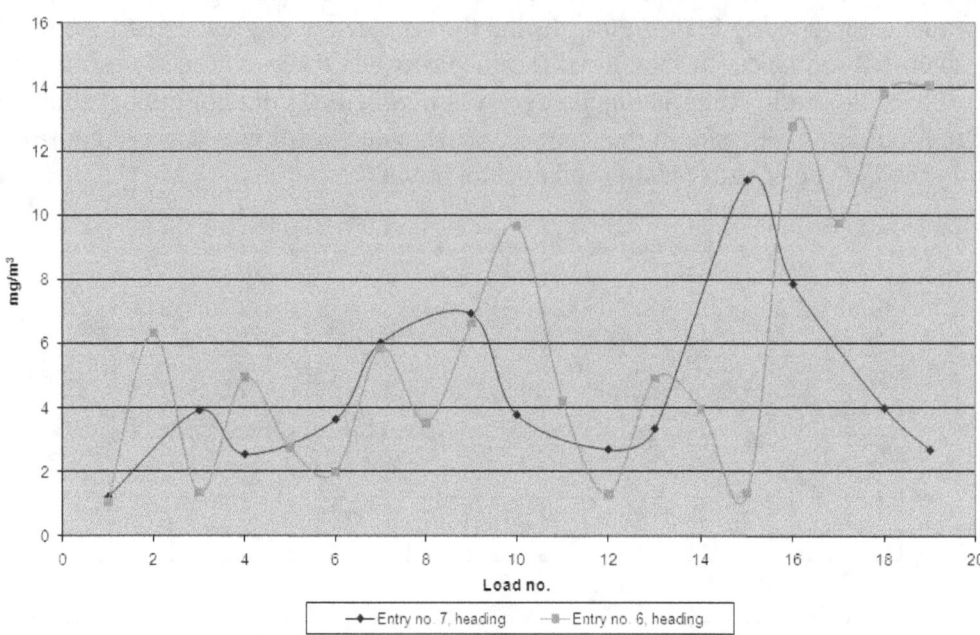

Figure A-4. Shuttle car dust exposures with blowing face ventilation when the miner was driving a heading.

Figure A-5 is a plot of shuttle car dust exposures for one cut under condition 3—continuous miner turning a right crosscut with blowing ventilation. When operating under these conditions, dust exposures in the shuttle cars decreased during loading in the deep-cut sequence. Dust levels averaged 5.26 mg/m^3 when loading during the regular-cut depth and 3.55 mg/m^3 during the deep-cut depth. The 85% CIs for these averages did not overlap, indicating a statistically significant difference.

During the study conducted at Mine A, the use of deep-cutting techniques had little bearing on the dust exposure levels of the shuttle car operators. When loading, transit, and the dumping sequences were included in exposure calculations, the shuttle cars averaged dust levels were 0.98 mg/m^3 during the regular-cut depth and 1.22 mg/m^3 during the deep-cut depth, which is not a statistically significant (85% CIs) difference. While the depth of cut did not affect dust levels, ventilation type had a very pronounced impact on the shuttle car operators' exposure levels. During production, the average shuttle car exposure was 0.22 mg/m^3 when using exhausting ventilation and 1.98 mg/m^3 when using blowing ventilation, a highly statistically significant (95% CIs) difference.

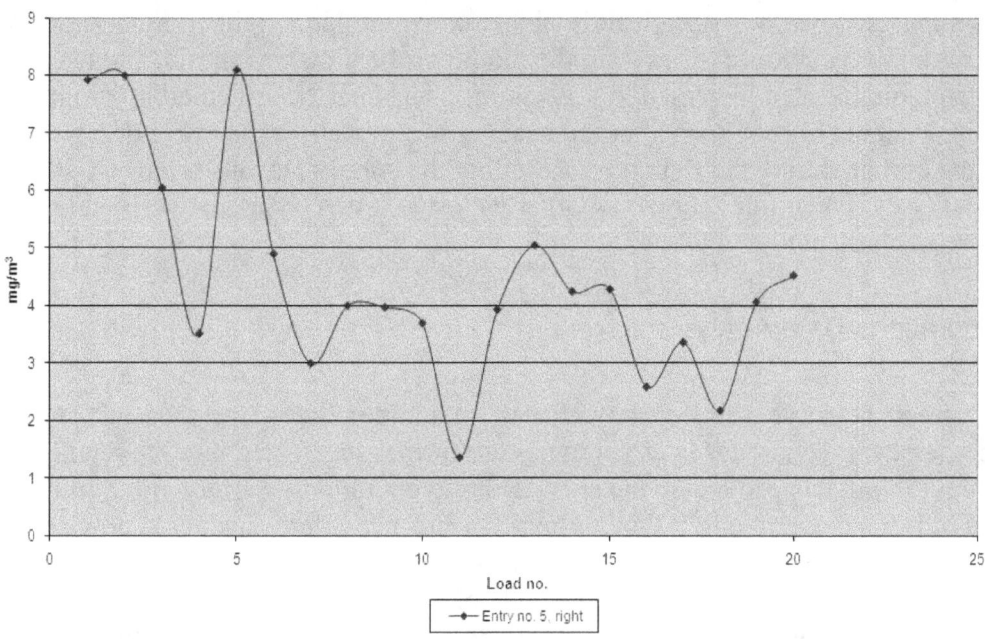

Figure A-5. Shuttle car dust exposures with blowing face ventilation when the miner was turning a right crosscut.

Miner-generated Dust

Table A-4 shows miner-generated dust levels during the regular- and deep-cut sequences for each room development. The last column of Table A-4 adjusts dust levels based on the measured productivity for each cut sequence using the normalization process described earlier in the "Sampling Protocol and Data Analysis" section. This process normalizes dust levels to productivity using the average number of cars loaded per minute for the section, which was observed to be 0.37 cars per minute for Mine A.

When using exhausting ventilation, miner-generated dust levels averaged 1.78 mg/m^3 during regular-cut depth and 2.07 mg/m^3 during deep-cut depth. When using blowing ventilation, miner-generated dust levels averaged 2.47 mg/m^3 during regular-cut depth and 2.46 mg/m^3 during deep-cut depth. The differences in these averages were not statistically significant (85% CIs), indicating that exposures downwind of the miner were unaffected by deep-cutting practices.

Miner Operator Data Analysis

Table A-5 shows miner operator exposure levels as measured with a PDM device during each cut of the study. The levels presented in the table isolate production time only and are not indicative of shift average levels, which were much lower. Dust exposures were adjusted for productivity. Depth of cut did not significantly (85% CIs) impact the miner operator's exposure levels for either ventilation type, averaging 0.80 mg/m^3 during regular-cut depth and 1.36 mg/m^3

during deep-cut depth with exhausting face ventilation, and 2.92 mg/m^3 during regular-cut depth and 2.05 mg/m^3 during deep-cut depth with blowing face ventilation. However, the miner operator's exposure level was significantly higher when turning a right crosscut. For this cut, the miner operator was positioned away from the mouth of the blowing curtain, and his average exposure to respirable dust increased to 5.17 mg/m^3. This may be remedied by dropping the curtain back a couple of roof bolts after the heading is cut to allow the operator to remain at the mouth of the curtain during the right turn. Resetting the curtain may also improve shuttle car exposures. The effects of this action on worker exposure levels should be verified using a real-time device such as a pDR or PDM.

Bolter Operations Data Analysis

Table A-6 shows bolter data collected at Mine A, including bolter-generated dust and left- and right-side operator exposure levels. Dust levels are further segmented into the regular- (0–20 ft) and deep-cut (20–40 ft) positions in the entry. Bolting operations in Entry No. 2 were conducted downwind of mining operations. All other bolting sequences were conducted upwind of the miner. During Day 1 of the study, both bolted entries were set up with exhausting curtain; however, no measureable airflow was detected behind the curtain. During Day 2, the entries were set up with blowing ventilation, and the airflow readings behind the curtain were 5,200 cfm for Entry No. 6 and 3,400 cfm for Entry No. 7. It is interesting to note that, when no airflow was measured, dust levels did appear to increase as the bolting sequence progressed from regular- to deep-cut depth, indicating a possible accumulation of dust due to low circulation of air. However, increases in dust levels were not observed on Day 2, when airflow behind the curtain was measureable. In addition, the dust exposure levels of the bolter operators, when operating downwind of the miner, were almost twice the levels experienced when operating upwind of the miner. Exposures when operating downwind of the miner were 3.00 mg/m^3 at the left-side position and 2.19 mg/m^3 at the right-side position. This mine minimized downwind bolting operations, and the shift average dust exposure levels of the bolter operators were low throughout the study. The left-side operator's shift average dust exposure level was 0.99 mg/m^3 on Day 1 of the study. Day 2 left-side data were lost due to a PDM battery failure. The right-side operator's shift average dust exposure levels were 1.06 mg/m^3 on Day 1 and 0.91 mg/m^3 on Day 2. The bolting machine's contribution to dust in the return air (bolter-generated dust) was not significant throughout the study compared to other sources, averaging less than 0.1 mg/m^3.

Dust-control Monitoring

Throughout the study, the line curtain and airflow to the face were well maintained. When using exhausting curtain, the average starting scrubber airflow was 11,500 cfm, while airflow delivered to the face averaged 11,500 cfm before activation of the scrubber and 16,700 cfm after activation. Similarly, when using blowing curtain, the average starting scrubber airflow was 10,800 cfm, while airflow behind the curtain was 9,600 cfm before activation of the scrubber and 14,800 after activation.

As mentioned earlier, one of the concerns was that the scrubber would become clogged with particles as the mining cycle progressed through the deep cut, diminishing its effectiveness. This condition was not observed at Mine A, as the average scrubber airflow dropped only 4% from start to finish during the cuts. The scrubber screen was tapped and back-flushed before each cut, a practice that should be continued to assure proper scrubber function. Likewise, no degradation in bolter suction pressures was observed during the study, as every reading registered 12 inches of mercury (-in Hg).

Summary

1. When driving a heading using exhausting face ventilation, shuttle car dust levels during loading did not appear to be affected by deep-cut practices.

2. When driving a heading using blowing face ventilation, shuttle car dust levels appeared to be higher during loading in the deep-cut sequence when compared to the regular-cut sequence.

3. When turning a right crosscut using blowing ventilation, shuttle car dust levels appeared to be lower during loading in the deep-cut sequence when compared to the regular-cut sequence.

4. Average shuttle car dust levels during the regular- and deep-cut sequences were similar when loading, transit, and dumping operations were included in the calculation.

5. When loading, transit, and dumping operations were included, dust levels in the shuttle cars were 1.76 mg/m^3 higher when using blowing ventilation versus exhausting ventilation. To minimize the dust exposure levels of the shuttle car operators, preference should be given to exhausting over blowing face ventilation when possible, particularly in Entry No. 4, which can be ventilated either way without interfering with shuttle car traffic.

6. Deep-cut practices did not influence miner-generated dust levels during the study.

7. Deep-cut practices did not influence miner operator exposure levels during the study.

8. During production, the miner operator's dust exposure levels increased by 4.10 mg/m^3 when turning a right crosscut as compared to driving a heading. During this cut, the miner operator was positioned away from the mouth of the blowing curtain. If the ventilation plan allows, this may be remedied by dropping the curtain back a couple of roof bolts after the heading is cut to allow the operator to remain at the mouth of the curtain during the right turn. Resetting the curtain may also improve the shuttle car operator exposure levels. The effects of this action on worker exposure levels should be verified using a real-time device such as a pDR or PDM.

9. The bolter operators' dust exposure levels increased as the bolting cycle progressed when airflow behind the curtain was not measureable. Airflow levels of 3,400 to 5,200 cfm were sufficient to clear the bolting face of accumulated dust.

10. Scrubber performance did not significantly degrade during the deep-cut sequence as scrubber airflow dropped by an average of 4% from start to completion of the cuts.

11. Bolter dust collector suction pressure did not significantly degrade due to deep-cutting practices and remained consistent at 12 in Hg for each measurement.

Table A-1. Shuttle car dust levels when loading at the face for Cuts No. 1, 2, and 3
Note: Shaded areas show deep-cut depth, and clear areas show regular-cut depth.
Airflows are before scrubber activation.

Cut No.	Day	Cut Sequence	Curtain Airflow (cfm)	Face Ventilation	Car No.	Time at Face	Shuttle Car Dust (mg/m³)	Intake Dust Level (mg/m³)	Adjusted Shuttle Car Dust (mg/m³)
1	1	No. 3, Heading	10,100	Exhausting	2	8:56:42 to 8:57:40	0.44	0.01	0.43
1	1	No. 3, Heading	10,100	Exhausting	3	9:01:20 to 9:02:30	No Data	No Data	No Data
1	1	No. 3, Heading	10,100	Exhausting	2	9:03:20 to 9:04:30	0.09	0.01	0.08
1	1	No. 3, Heading	10,100	Exhausting	3	9:06:40 to 9:08:04	No Data	No Data	No Data
1	1	No. 3, Heading	10,100	Exhausting	1	9:08:20 to 9:09:40	0.05	0.00	0.05
1	1	No. 3, Heading	10,100	Exhausting	2	9:09:40 to 9:11:24	0.22	0.02	0.20
1	1	No. 3, Heading	10,100	Exhausting	3	9:12:30 to 9:14:50	No Data	No Data	No Data
1	1	No. 3, Heading	10,100	Exhausting	2	9:15:40 to 9:17:14	0.15	0.01	0.14
1	1	No. 3, Heading	10,100	Exhausting	1	9:17:30 to 9:19:00	0.05	0.00	0.05
1	1	No. 3, Heading	10,100	Exhausting	3	9:19:20 to 9:21:03	No Data	No Data	No Data
1	1	No. 3, Heading	10,100	Exhausting	2	9:21:45 to 9:23:58	0.02	0.01	0.01
1	1	No. 3, Heading	10,100	Exhausting	1	9:24:18 to 9:28:05	0.02	0.00	0.02
1	1	No. 3, Heading	10,100	Exhausting	3	9:28:27 to 9:31:48	No Data	No Data	No Data
1	1	No. 3, Heading	10,100	Exhausting	1	9:32:10 to 9:34:20	0.73	0.01	0.73
1	1	No. 3, Heading	10,100	Exhausting	2	9:34:40 to 9:36:32	0.13	0.01	0.12
1	1	No. 3, Heading	10,100	Exhausting	3	9:37:38 to 9:39:28	No Data	No Data	No Data
1	1	No. 3, Heading	10,100	Exhausting	1	9:39:36 to 9:41:54	0.08	0.01	0.07
1	1	No. 3, Heading	10,100	Exhausting	2	9:42:07 to 9:43:40	0.74	0.02	0.72
1	1	No. 3, Heading	10,100	Exhausting	3	9:44:30 to 9:46:10	No Data	No Data	No Data
2	1	No. 2, Heading	13,500	Exhausting	1	10:09:24 to 10:11:11	0.09	0.03	0.06
2	1	No. 2, Heading	13,500	Exhausting	3	10:11:40 to 10:13:06	No Data	No Data	No Data
2	1	No. 2, Heading	13,500	Exhausting	2	10:14:24 to 10:15:58	1.65	0.05	1.60
2	1	No. 2, Heading	13,500	Exhausting	1	10:16:50 to 10:18:35	0.14	0.01	0.13
2	1	No. 2, Heading	13,500	Exhausting	3	10:23:00 to 10:24:30	No Data	No Data	No Data
2	1	No. 2, Heading	13,500	Exhausting	1	10:24:45 to 10:26:18	0.12	0.00	0.11
2	1	No. 2, Heading	13,500	Exhausting	2	10:26:40 to 10:27:56	0.30	0.05	0.25
2	1	No. 2, Heading	13,500	Exhausting	3	10:30:01 to 10:32:16	No Data	No Data	No Data
2	1	No. 2, Heading	13,500	Exhausting	1	10:32:37 to 10:35:40	0.12	0.02	0.10
2	1	No. 2, Heading	13,500	Exhausting	2	10:35:55 to 10:37:29	0.19	0.07	0.13
2	1	No. 2, Heading	13,500	Exhausting	3	10:39:10 to 10:40:48	No Data	No Data	No Data
2	1	No. 2, Heading	13,500	Exhausting	1	10:41:08 to 10:42:43	0.46	0.01	0.45
2	1	No. 2, Heading	13,500	Exhausting	2	10:42:58 to 10:44:34	0.86	0.05	0.81
2	1	No. 2, Heading	13,500	Exhausting	3	10:46:22 to 10:48:32	No Data	No Data	No Data
2	1	No. 2, Heading	13,500	Exhausting	1	10:48:47 to 10:50:29	3.67	0.02	3.65
2	1	No. 2, Heading	13,500	Exhausting	2	10:50:50 to 10:52:10	0.69	0.13	0.57
2	1	No. 2, Heading	13,500	Exhausting	3	10:54:39 to 10:59:30	No Data	No Data	No Data
2	1	No. 2, Heading	13,500	Exhausting	1	11:02:02 to 11:03:33	0.66	0.01	0.65
3	1	No. 1, Heading	10,900	Exhausting	2	11:36:03 to 11:37:23	0.03	0.01	0.01
3	1	No. 1, Heading	10,900	Exhausting	1	11:38:42 to 11:40:00	0.09	0.02	0.08
3	1	No. 1, Heading	10,900	Exhausting	3	11:40:48 to 11:42:19	No Data	No Data	No Data
3	1	No. 1, Heading	10,900	Exhausting	1	11:44:08 to 11:45:33	0.16	0.06	0.10
3	1	No. 1, Heading	10,900	Exhausting	2	11:46:09 to 11:47:21	0.11	0.07	0.04
3	1	No. 1, Heading	10,900	Exhausting	1	11:49:45 to 11:51:13	0.14	0.06	0.08
3	1	No. 1, Heading	10,900	Exhausting	3	11:52:19 to 11:53:38	No Data	No Data	No Data
3	1	No. 1, Heading	10,900	Exhausting	1	11:55:50 to 11:56:47	0.15	0.04	0.11
3	1	No. 1, Heading	10,900	Exhausting	2	11:57:22 to 11:58:18	0.15	0.14	0.01
3	1	No. 1, Heading	10,900	Exhausting	3	12:00:46 to 12:02:25	No Data	No Data	No Data
3	1	No. 1, Heading	10,900	Exhausting	1	12:03:19 to 12:04:51	0.11	0.03	0.08
3	1	No. 1, Heading	10,900	Exhausting	2	12:05:53 to 12:07:21	0.14	0.09	0.06
3	1	No. 1, Heading	10,900	Exhausting	1	12:09:06 to 12:10:53	0.08	0.02	0.06
3	1	No. 1, Heading	10,900	Exhausting	3	12:12:11 to 12:13:46	No Data	No Data	No Data
3	1	No. 1, Heading	10,900	Exhausting	2	12:16:05 to 12:17:28	0.37	0.07	0.30
3	1	No. 1, Heading	10,900	Exhausting	1	12:19:42 to 12:20:40	0.24	0.03	0.21
3	1	No. 1, Heading	10,900	Exhausting	3	12:21:40 to 12:24:12	No Data	No Data	No Data
3	1	No. 1, Heading	10,900	Exhausting	1	12:25:18 to 12:26:22	1.36	0.01	1.36

Table A-2. Shuttle car dust levels when loading at the face for Cuts No. 4, 5, and 6

Note: Shaded areas show deep-cut depth, and clear areas show regular-cut depth. Airflows are after scrubber activation.

Cut No.	Day	Cut Sequence	Curtain Airflow (cfm)	Face Ventilation	Car No.	Time at Face	Shuttle Car Dust (mg/m^3)	Intake Dust Level (mg/m^3)	Adjusted Shuttle Car Dust (mg/m^3)
4	1	No. 7, Heading	18,000	Blowing	1	13:06:05 to 13:07:21	1.22	0.00	1.22
4	1	No. 7, Heading	18,000	Blowing	3	13:09:47 to 13:11:09	No Data	No Data	No Data
4	1	No. 7, Heading	18,000	Blowing	1	13:12:10 to 13:13:23	3.92	0.00	3.92
4	1	No. 7, Heading	18,000	Blowing	2	13:14:13 to 13:15:20	2.56	0.00	2.56
4	1	No. 7, Heading	18,000	Blowing	3	13:17:20 to 13:19:22	No Data	No Data	No Data
4	1	No. 7, Heading	18,000	Blowing	1	13:21:05 to 13:23:43	3.64	0.00	3.64
4	1	No. 7, Heading	18,000	Blowing	2	13:24:30 to 13:25:27	6.02	0.00	6.02
4	1	No. 7, Heading	18,000	Blowing	3	13:27:30 to 13:29:00	No Data	No Data	No Data
4	1	No. 7, Heading	18,000	Blowing	1	13:30:16 to 13:31:09	6.94	0.00	6.94
4	1	No. 7, Heading	18,000	Blowing	2	13:32:01 to 13:33:20	3.78	0.00	3.78
4	1	No. 7, Heading	18,000	Blowing	3	13:35:14 to 13:36:35	No Data	No Data	No Data
4	1	No. 7, Heading	18,000	Blowing	1	13:37:49 to 13:39:09	2.70	0.00	2.70
4	1	No. 7, Heading	18,000	Blowing	2	13:40:05 to 13:41:15	3.35	0.00	3.35
4	1	No. 7, Heading	18,000	Blowing	3	13:43:07 to 13:44:21	No Data	No Data	No Data
4	1	No. 7, Heading	18,000	Blowing	1	13:45:36 to 13:46:48	11.10	0.00	11.10
4	1	No. 7, Heading	18,000	Blowing	2	13:47:51 to 13:49:00	7.87	0.00	7.87
4	1	No. 7, Heading	18,000	Blowing	3	13:51:14 to 13:52:50	No Data	No Data	No Data
4	1	No. 7, Heading	18,000	Blowing	1	13:53:50 to 13:55:21	3.98	0.00	3.98
4	1	No. 7, Heading	18,000	Blowing	2	13:56:19 to 13:57:36	2.68	0.00	2.68
5	2	No. 6, Heading	13,000	Blowing	1	10:14:43 to 10:16:16	1.063	0.02	1.04
5	2	No. 6, Heading	13,000	Blowing	2	10:17:21 to 10:18:25	6.326	0.03	6.29
5	2	No. 6, Heading	13,000	Blowing	1	10:30:46 to 10:31:53	1.361	0.02	1.35
5	2	No. 6, Heading	13,000	Blowing	2	10:33:02 to 10:34:05	4.934	0.04	4.89
5	2	No. 6, Heading	13,000	Blowing	1	10:35:50 to 10:37:35	2.751	0.01	2.74
5	2	No. 6, Heading	13,000	Blowing	2	10:38:47 to 10:40:42	1.995	0.02	1.98
5	2	No. 6, Heading	13,000	Blowing	1	10:41:54 to 10:42:54	5.846	0.01	5.84
5	2	No. 6, Heading	13,000	Blowing	2	10:44:37 to 10:45:36	3.530	0.06	3.47
5	2	No. 6, Heading	13,000	Blowing	1	10:46:55 to 10:48:01	6.626	0.03	6.59
5	2	No. 6, Heading	13,000	Blowing	2	10:49:13 to 10:50:27	9.652	0.03	9.62
5	2	No. 6, Heading	13,000	Blowing	1	10:52:10 to 10:53:54	4.201	0.01	4.19
5	2	No. 6, Heading	13,000	Blowing	2	11:01:30 to 11:03:05	1.304	0.00	1.30
5	2	No. 6, Heading	13,000	Blowing	1	11:04:25 to 11:05:46	4.918	0.00	4.92
5	2	No. 6, Heading	13,000	Blowing	2	11:07:05 to 11:08:15	3.962	0.00	3.96
5	2	No. 6, Heading	13,000	Blowing	1	11:10:08 to 11:13:06	1.336	0.00	1.34
5	2	No. 6, Heading	13,000	Blowing	2	11:14:21 to 11:15:40	12.769	0.00	12.77
5	2	No. 6, Heading	13,000	Blowing	1	11:17:15 to 11:18:23	9.741	0.00	9.74
5	2	No. 6, Heading	13,000	Blowing	2	11:19:50 to 11:21:03	13.791	0.00	13.79
5	2	No. 6, Heading	13,000	Blowing	1	11:22:28 to 11:23:30	14.035	0.00	14.04
6	2	No. 5, Heading	14,000	Blowing	1	11:50:37 to 11:52:11	0.792	0.03	0.77
6	2	No. 5, Heading	14,000	Blowing	2	11:54:05 to 11:55:20	4.120	0.02	4.10
6	2	No. 5, Heading	14,000	Blowing	1	11:56:56 to 11:58:20	0.510	0.00	0.51
6	2	No. 5, Heading	14,000	Blowing	2	11:59:44 to 12:01:15	5.025	0.01	5.01
6	2	No. 5, Heading	14,000	Blowing	1	12:02:25 to 12:04:23	2.213	0.31	1.90
6	2	No. 5, Heading	14,000	Blowing	2	12:05:49 to 12:07:00	5.820	0.14	5.68
6	2	No. 5, Heading	14,000	Blowing	1	12:08:02 to 12:09:26	3.680	0.31	3.37
6	2	No. 5, Heading	14,000	Blowing	2	12:10:51 to 12:11:55	5.597	0.08	5.52
6	2	No. 5, Heading	14,000	Blowing	1	12:13:22 to 12:14:14	7.087	0.01	7.08
6	2	No. 5, Heading	14,000	Blowing	2	12:15:40 to 12:17:00	6.461	0.00	6.46
6	2	No. 5, Heading	14,000	Blowing	1	12:18:31 to 12:21:25	7.005	0.15	6.86

Table A-3. Shuttle car dust levels when loading at the face for Cut No. 7
Note: Shaded areas show deep-cut depth, and clear areas show regular-cut depth.
Airflows are after scrubber activation.

Cut No.	Day	Cut Sequence	Curtain Airflow (cfm)	Face Ventilation	Car No.	Time at Face	Shuttle Car Dust (mg/m³)	Intake Dust Level (mg/m³)	Adjusted Shuttle Car Dust (mg/m³)
7	2	No. 5, Right	14,000	Blowing	2	12:31:53 to 12:34:08	7.92	0.07	7.85
7	2	No. 5, Right	14,000	Blowing	1	12:35:24 to 12:36:22	8.00	0.02	7.98
7	2	No. 5, Right	14,000	Blowing	2	12:37:46 to 12:39:10	6.04	0.14	5.90
7	2	No. 5, Right	14,000	Blowing	1	12:40:30 to 12:42:25	3.50	0.02	3.48
7	2	No. 5, Right	14,000	Blowing	2	12:43:45 to 12:44:46	8.10	0.07	8.03
7	2	No. 5, Right	14,000	Blowing	1	12:46:30 to 12:47:35	4.90	0.06	4.84
7	2	No. 5, Right	14,000	Blowing	2	12:49:30 to 12:52:30	2.98	0.08	2.90
7	2	No. 5, Right	14,000	Blowing	1	12:53:55 to 12:55:03	3.99	0.03	3.96
7	2	No. 5, Right	14,000	Blowing	2	12:56:22 to 12:57:35	3.97	0.00	3.96
7	2	No. 5, Right	14,000	Blowing	1	12:58:54 to 13:01:00	3.68	0.00	3.68
7	2	No. 5, Right	14,000	Blowing	2	13:02:16 to 13:06:27	1.34	0.00	1.34
7	2	No. 5, Right	14,000	Blowing	1	13:08:10 to 13:09:30	3.93	0.00	3.93
7	2	No. 5, Right	14,000	Blowing	2	13:11:00 to 13:12:05	5.05	0.00	5.05
7	2	No. 5, Right	14,000	Blowing	1	13:15:04 to 13:16:38	4.24	0.00	4.24
7	2	No. 5, Right	14,000	Blowing	2	13:18:06 to 13:19:10	4.28	0.00	4.28
7	2	No. 5, Right	14,000	Blowing	1	13:20:41 to 13:22:08	2.58	0.00	2.58
7	2	No. 5, Right	14,000	Blowing	2	13:23:36 to 13:26:10	3.35	0.00	3.35
7	2	No. 5, Right	14,000	Blowing	1	13:28:00 to 13:29:40	2.17	0.00	2.17
7	2	No. 5, Right	14,000	Blowing	2	13:31:01 to 13:32:03	4.06	0.00	4.06
7	2	No. 5, Right	14,000	Blowing	1	13:33:33 to 13:34:47	4.52	0.00	4.52

Table A-4. Continuous miner-generated dust during production as measured with pDRs calibrated with gravimetric samplers
Note: Airflows are before scrubber activation.

Cut No.	Cut Sequence	Depth	Curtain Airflow (cfm)	Face Ventilation	Starting Scrubber Airflow (cfm)	Ending Scrubber Airflow (cfm)	Miner Intake Dust Level (mg/m³)	Miner Return Dust Level (mg/m³)	Miner Generated (mg/m³)	Cars Per Minute	Adjusted Miner Generated (mg/m³)
1	No. 3, Heading	Regular	10,100	Exhausting	11,100	N/A	0.01	3.08	3.07	0.41	2.77
1	No. 3, Heading	Deep	10,100	Exhausting	N/A	10,000	0.01	1.70	1.69	0.37	1.69
2	No. 2, Heading	Regular	13,500	Exhausting	11,700	N/A	0.03	1.44	1.41	0.34	1.53
2	No. 2, Heading	Deep	13,500	Exhausting	N/A	11,700	0.04	1.97	1.93	0.33	2.16
3	No. 1, Heading	Regular	10,900	Exhausting	11,700	N/A	0.06	1.19	1.12	0.40	1.04
3	No. 1, Heading	Deep	10,900	Exhausting	N/A	11,100	0.05	2.29	2.24	0.35	2.37
4	No. 7, Heading	Regular	10,100	Blowing	11,100	N/A	0.00	1.88	1.88	0.36	1.93
4	No. 7, Heading	Deep	10,100	Blowing	N/A	11,100	0.00	1.68	1.68	0.39	1.60
5	No. 6, Heading	Regular	9,500	Blowing	10,700	N/A	0.03	2.98	2.95	0.43	2.54
5	No. 6, Heading	Deep	9,500	Blowing	N/A	10,700	0.00	4.14	4.14	0.38	4.03
7	No. 5, Right	Regular	9,300	Blowing	10,700	N/A	0.05	2.76	2.71	0.34	2.95
7	No. 5, Right	Deep	9,300	Blowing	N/A	9,500	0.00	1.46	1.46	0.31	1.75

Table A-5. Miner operator exposures during production as measured with a PDM instrument

Note: Airflows are before scrubber activation.

Cut No.	Cut Sequence	Depth	Curtain Airflow (cfm)	Face Ventilation	Starting Scrubber Airflow (cfm)	Ending Scrubber Airflow (cfm)	Miner Operator Exposure (mg/m³)	Cars Per Minute	Adj. Miner Operator Exposure (mg/m³)
1	No. 3, heading	Regular	10,100	Exhausting	11,100	N/A	0.43	0.41	0.39
1	No. 3, heading	Deep	10,100	Exhausting	N/A	10,000	0.91	0.37	0.91
2	No. 2, heading	Regular	13,500	Exhausting	11,700	N/A	1.44	0.34	1.57
2	No. 2, heading	Deep	13,500	Exhausting	N/A	11,700	1.57	0.33	1.76
3	No. 1, heading	Regular	10,900	Exhausting	11,700	N/A	0.46	0.40	0.43
3	No. 1, heading	Deep	10,900	Exhausting	N/A	11,100	1.32	0.35	1.40
4	No. 7, heading	Regular	10,100	Blowing	11,100	N/A	0.65	0.36	0.67
4	No. 7, heading	Deep	10,100	Blowing	N/A	11,100	1.10	0.39	1.04
5	No. 6, heading	Regular	9,500	Blowing	10,700	N/A	1.69	0.43	1.45
5	No. 6, heading	Deep	9,500	Blowing	N/A	10,700	1.45	0.38	1.41
7	No. 5, right	Regular	9,300	Blowing	10,700	N/A	6.10	0.34	6.64
7	No. 5, right	Deep	9,300	Blowing	N/A	9,500	3.10	0.31	3.70

Table A-6. Bolter-generated dust and bolter operator exposures during the bolting cycle

Room No.	Entry	Depth	Position With Respect To Miner	Curtain Airflow (cfm)	Face Ventilation	Bolter Suction Left/Right (in Hg)	Left-Side Bolter Dust Levels (mg/m³)	Right-Side Bolter Dust Levels (mg/m³)	Upwind Bolter Dust Levels (mg/m³)	Downwind Bolter Dust Levels (mg/m³)	Bolter Generated Dust (mg/m³)
1	No. 3, heading	Regular	Upwind	None	Exhausting	12/12	0.65	0.96	0.02	0.07	0.05
1	No. 3, heading	Deep	Upwind	None	Exhausting	12/12	2.65	1.61	0.01	0.16	0.15
2	No. 2, heading	Regular	Downwind	None	Exhausting	12/12	2.28	1.35	0.28	0.24	0.00
2	No. 2, heading	Deep	Downwind	None	Exhausting	12/12	3.71	3.02	1.19	1.24	0.05
3	No. 7, heading	Regular	Upwind	3,400	Blowing	12/12	No Data	1.39	0.00	0.04	0.04
3	No. 7, heading	Deep	Upwind	3,400	Blowing	12/12	No Data	1.70	0.00	0.06	0.06
4	No. 6, heading	Regular	Upwind	5,200	Blowing	12/12	No Data	0.91	0.00	0.13	0.13
4	No. 6, heading	Deep	Upwind	5,200	Blowing	12/12	No Data	0.74	0.00	0.11	0.11

Appendix B: MINE B CASE STUDY

Mine-specific Information

Mine B used two Joy 14CM15 miners on this section. The miner spray configuration is shown in Figure B-1. A total of 30 dust suppression sprays were operated at a pressure of not less than 75 psi as required by the mine ventilation plan—19 sprays were located at the top of the ripper, 8 sprays were located at the top of the pan, and 3 sprays were located at the throat of the chain conveyor. The ventilation plan required the scrubber capacity to be a minimum of 4,000 cfm. The plan also required the line curtain to be maintained within 50 ft of the deepest point of penetration with the minimum required airflow from the curtain to be 4000 cfm with the scrubber on and 3600 cfm with the scrubber off.

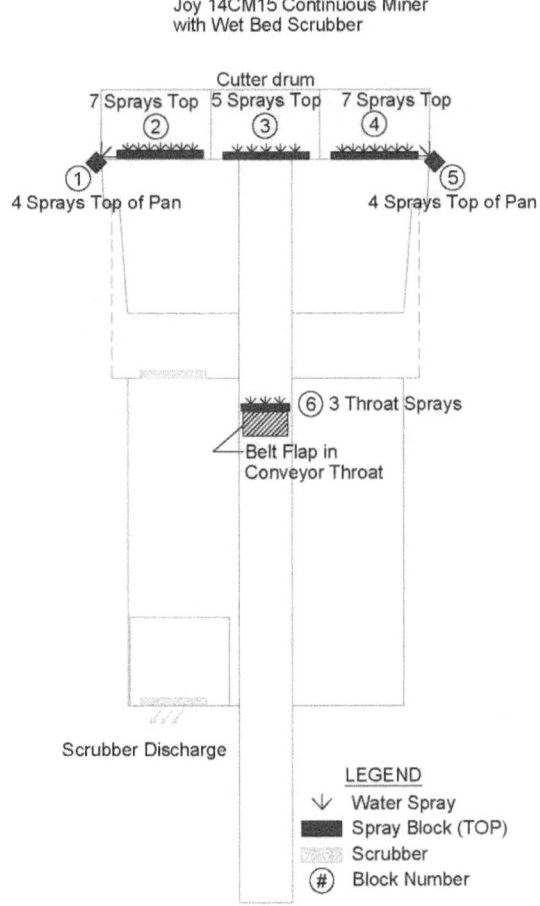

Figure B-1. Mine B continuous miner spray configuration.

The cut sequence for Mine B was as follows:

(1) 20-ft sump cut, left side
(2) 20-ft slab cut, right side

(3) 20-ft sump cut, right side
(4) 20-ft slab cut, left side

Figure B-2 shows a plan view of the Mine B at the time of this study. The mine used a split ventilation configuration with Entry No. 5 serving as the primary intake and Entries No. 1 and No. 8 serving as the returns. All other entries were neutral, including the beltway, which was located in Entry No. 4. Mining height varied between 54-in and 78-in with entry width averaging 20 ft. In Entry No. 3, the floor rock narrowed enough (17-in) to allow extraction of a lower and thinner (13-in) coal seam, resulting in a 78-in mining height in that entry. Pillar dimensions were 80 ft by 60 ft. The deep-cut depths were 40 ft.

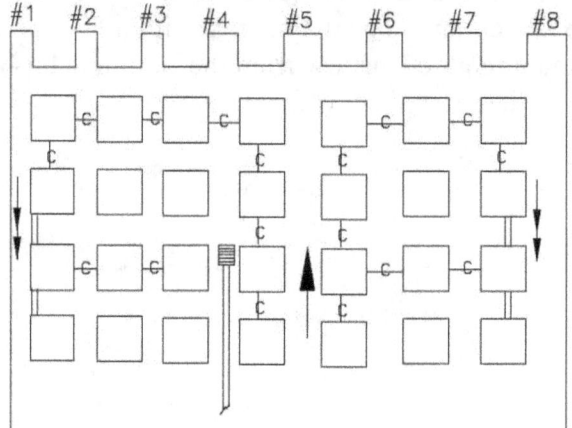

Figure B-2. Plan view of Mine B section.

The Mine B dust survey was conducted on April 14 and 16, 2009. April 15 was a mine-scheduled down-production day.

Shuttle Car Data Analysis

Tables B-1 and B-2 show the shuttle car data collected during the study. Sampling was only conducted on the miner located on the left side of the section, which was responsible for mining Entries No. 1, 2, 3, and 4. The miner on the right side was not sampled. Three unique types of deep cuts were evaluated for this study:

(1) Continuous miner driving a heading with exhausting curtain
(2) Continuous miner driving a heading without curtain
(3) Continuous miner turning a left crosscut without curtain

Headings and left crosscuts represented the only types of cuts used on the left side of the section as all crosscuts were turned toward return airflow.

Figure B-3 is a plot of shuttle car dust exposures for three cuts when the miner was driving a heading with exhausting curtain. The x-axis corresponds to the sequential load number during the cut. As seen in the plot, the number of cars for each deep cut varied between entries due to the fact that mining height also varied between entries.

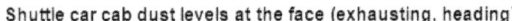
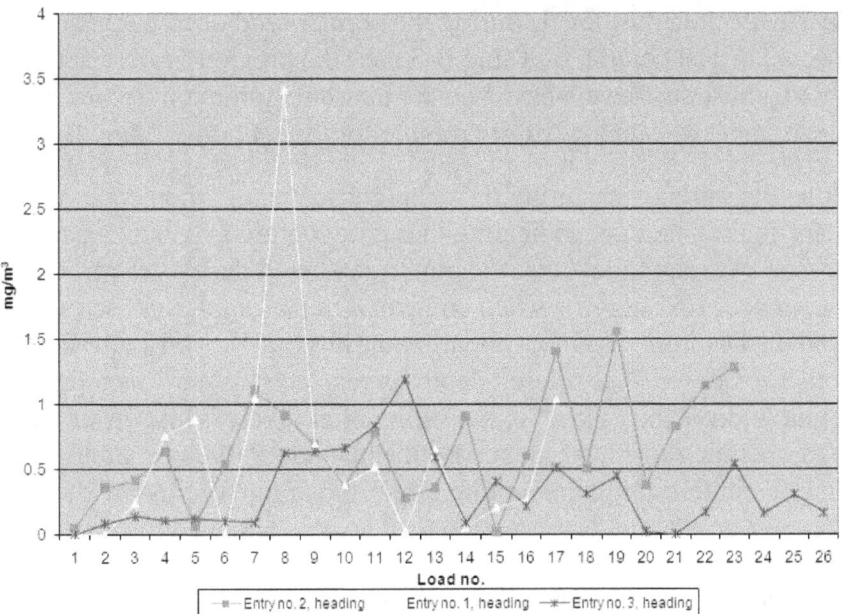

Figure B-3. Shuttle car dust levels when miner was driving a straight heading.
The lack of any apparent trend in Figure B-3 indicates that cutting depth did not affect the dust exposure levels of the shuttle car operators when the miner was driving a heading with exhausting face ventilation. During Cuts No. 2, 3, and 5, the average dust exposures in the cabs attributable to face operations were 0.54 mg/m^3 for the regular-cut sequence and 0.48 mg/m^3 for the deep-cut sequence. This difference was not statistically significant when comparing the 85% CIs for the averages.

Similar results were observed for the cuts completed without the use of ventilation curtain. Ventilation curtain was not needed for several of the cuts because of the variance allowing the curtain to be maintained within 50 ft of the deepest point of penetration. When driving a heading without curtain, dust levels in the cabs were 0.23 mg/m^3 during the regular-cut depth and 0.36 mg/m^3 during the deep-cut depth. When turning a left crosscut without curtain, dust levels were 0.25 mg/m^3 for the regular-cut depth and 0.12 mg/m^3 for the deep-cut depth. Each of these comparisons was not statistically significant (85% CIs).

An analysis of individual cuts found no significant change in dust levels when comparing the regular- to the deep-cut depth. No differences were found despite the fact that the mine experienced a wide range of scrubber airflows and face ventilation conditions between cuts. Table B-3 summarizes these conditions:

(1) Cuts without exhausting curtain
(2) Cuts using exhausting curtain; airflows measured behind the curtain before activation of the scrubber ranged from 5,130 to 9,000 cfm.
(3) Starting scrubber airflows ranging from 7,020 to 7,940 cfm
(4) Ending scrubber airflows ranging from 4,820 to 7,500 cfm

Miner-generated Dust

Table B-3 shows miner-generated dust levels during the regular- and deep-cut sequences for each room development. The last column of Table B-3 uses the previously described normalization process to adjust dust levels based on the measured productivity for each cut sequence. The average number of cars loaded per minute for the section at Mine B was 0.31.

Miner-generated dust levels were lower during the deep-cut sequence when compared to the regular-cut sequence for four of six cuts, indicating that dust capture may have been improving as mining depth increased. Averaging all cuts, the miner-generated dust level was 66% lower during the deep-cut sequence (1.52 mg/m^3) when compared to the regular-cut sequence (4.51 mg/m^3). This improved dust capture was most apparent during Cut No. 4. Figure B-4 shows miner return dust levels for Cut No. 4 as mining depth increased. This cut was unique because the miner had just begun to drive the Entry No. 4 heading after all crosscut entries had been opened. The miner was operating just inby the crosscut and over 19,000 cfm of air was sweeping across the miner during the initial sump, preventing effective capture of dust by the scrubber. It is interesting to note, however, that miner-generated dust levels continued to decrease as mining depth increased. This may be due to the fact that airflow reaching the face is essentially limited to what the scrubber is drawing after the miner extends beyond the influence of curtain or entry ventilation, minimizing dust blowby at the mouth of the scrubber. For all other heading cuts observed during the study, the miner was sumped into the coal by at least 10 ft before the cut began, preventing the ventilation air from sweeping over the head of the miner at the beginning of the cut. Eliminating the unique Cut No. 4 from the calculation indicates that miner-generated dust is reduced by an average of 28% during the deep-cut depth. Cut No. 3 was the only heading cut that experienced increased miner-generated dust levels as the cutting depth increased. This may be explained by the fact that scrubber airflow performance degraded by the largest percentage (35%) during Cut No. 3 when compared to all other cuts. Miner-generated dust also appeared to be somewhat higher during the deep-cut sequence for Cut No. 6, when the miner was driving a left crosscut.

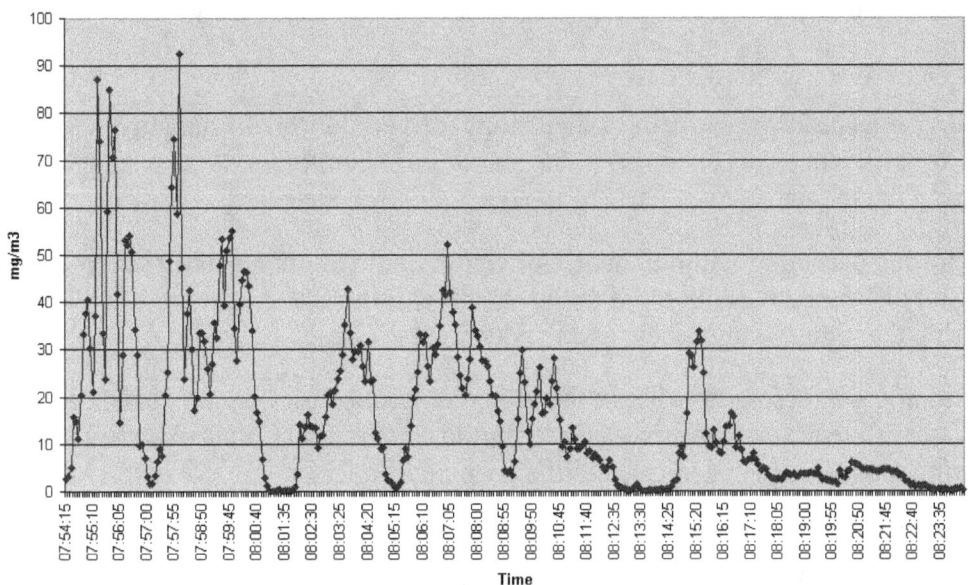

Figure B-4. Miner return dust levels during Cut No. 4.

Miner Operator Data Analysis

Table B-4 shows the miner operator exposure levels as measured with a PDM device during each cut of the study. The last column of Table B-4 shows levels normalized for productivity. Depth of cut appeared to have no bearing on the miner operator's dust exposure level. The operator's exposure level was very low for both the regular-cut and deep-cut sequences, averaging 0.66 mg/m^3 for the regular-cut sequence and 0.45 mg/m^3 for the deep-cut sequence. In general, the operator maintained his position on the off-curtain side of the miner, parallel to or slightly outby the mouth of the exhaust curtain.

Bolter Operations Data Analysis

Table B-5 shows bolter operator dust exposure levels during the regular- and deep-cut sequences for each room that was bolted. Exposures were determined from PDM devices worn by the operators. Only Rooms No. 2 and 3 were conducted upwind of mining operations. For these rooms, there were no discernable differences in dust levels when comparing the regular- to deep-cut depth. This may be explained by the fact that throughout the study there was very little drop-off in the suction pressure of the bolting machine's dust collection system when comparing the start of the bolting cycle to the end of the cycle. Suction pressures did vary somewhat from day to day and when comparing the left side of the machine to the right side; however, measurements were always between 12 and 17 in Hg, which appeared to be sufficient for controlling dust. Likewise, the line curtain was well maintained and provided adequate airflow (2,920 to 4,680 cfm) to each bolting face. Dust exposures on the bolting faces increased by approximately 0.4 mg/m^3 when bolting was conducted downwind of mining operations during the study. In general, daily-average bolter operator dust exposure levels were quite low throughout the 2-day study, falling between 0.30 and 0.82 mg/m^3.

Summary

1. When driving a heading and using exhausting face ventilation, no statistically significant (85% CIs) difference was observed in the shuttle car operators' dust exposure levels when comparing the regular-cut sequence to the deep-cut sequence.

2. When driving a heading without the use of ventilation curtain, no statistically significant (85% CIs) difference was observed in the shuttle car operators' dust exposure levels when comparing the regular-cut sequence to the deep-cut sequence.

3. When turning a left crosscut without the use of ventilation curtain, no statistically significant (85% CIs) difference was observed in the shuttle car operators' dust exposure levels when comparing the regular-cut sequence to the deep-cut sequence.

4. Averaging all cuts, the miner-generated dust level was 66% lower during the deep-cut sequence when compared to the regular-cut sequence. This may be due to the fact that airflow reaching the face is essentially limited to what the scrubber is drawing after the miner extends beyond the influence of curtain or entry ventilation, maximizing dust capture at the mouth of the scrubber. Eliminating the unique Cut No. 4 from the calculation indicates that miner-generated dust is reduced by an average of 28% during the deep-cut depth.

5. Cut No. 3 was the only heading cut that experienced increased miner-generated dust levels as the cutting depth increased. This may be explained by the fact that scrubber airflow performance degraded by the largest percentage (35%) during Cut No. 3 when compared to all other cuts.

6. Miner-generated dust was somewhat higher during the deep-cut sequence for Cut No. 6, when the miner was driving a left crosscut.

7. Depth of cut had no bearing on the miner operator's dust exposure levels.

8. The bolter operators' dust exposure levels did not increase as the bolting cycle progressed into the deep-cut depth.

Table B-1. Shuttle car dust levels when loading at the face during Cuts No. 1 through 3
Note: Shaded areas show deep-cut depth, and clear areas show regular-cut depth. Airflows are before scrubber activation.

Cut No.	Day	Cut Sequence	Curtain Airflow (cfm)	Face Ventilation	Car No.	Time at Face	Shuttle Car Dust (mg/m^3)	Intake Dust Level (mg/m^3)	Adjusted Shuttle Car Dust (mg/m^3)
1	1	No. 1, Heading	0	No Curtain	1	8:15:13 to 8:16:20	0.16	0.15	0.01
1	1	No. 1, Heading	0	No Curtain	1	8:19:10 to 8:20:28	0.15	0.07	0.08
1	1	No. 1, Heading	0	No Curtain	1	8:23:00 to 8:25:15	0.12	0.10	0.02
1	1	No. 1, Heading	0	No Curtain	1	8:28:15 to 8:29:45	0.13	0.06	0.06
1	1	No. 1, Heading	0	No Curtain	1	8:32:05 to 8:33:22	0.32	0.16	0.16
1	1	No. 1, Heading	0	No Curtain	1	8:35:37 to 8:38:25	0.09	0.03	0.06
1	1	No. 1, Heading	0	No Curtain	1	8:40:45 to 8:42:23	0.10	0.06	0.05
1	1	No. 1, Heading	0	No Curtain	1	8:45:15 to 8:46:50	0.04	0.01	0.03
1	1	No. 1, Heading	0	No Curtain	1	8:51:40 to 8:54:40	0.07	0.04	0.03
1	1	No. 1, Heading	0	No Curtain	1	8:56:57 to 8:58:18	0.20	0.10	0.10
1	1	No. 1, Heading	0	No Curtain	1	9:00:58 to 9:02:15	0.15	0.14	0.01
1	1	No. 1, Heading	0	No Curtain	1	9:05:28 to 9:06:50	0.13	0.08	0.05
1	1	No. 1, Heading	0	No Curtain	1	9:09:30 to 9:10:50	0.14	0.05	0.09
1	1	No. 1, Heading	0	No Curtain	1	9:13:00 to 9:14:55	0.14	0.09	0.05
1	1	No. 1, Heading	0	No Curtain	2	9:19:30 to 9:21:00	0.12	0.12	0.00
2	1	No. 2, Heading	8,100	Exhausting	1	10:41:10 to 10:42:25	0.10	0.05	0.05
2	1	No. 2, Heading	8,100	Exhausting	2	10:43:10 to 10:44:25	0.41	0.05	0.36
2	1	No. 2, Heading	8,100	Exhausting	1	10:45:20 to 10:47:07	0.46	0.05	0.41
2	1	No. 2, Heading	8,100	Exhausting	2	10:47:20 to 10:48:50	0.70	0.07	0.63
2	1	No. 2, Heading	8,100	Exhausting	1	10:49:35 to 10:51:05	0.08	0.02	0.06
2	1	No. 2, Heading	8,100	Exhausting	2	10:52:05 to 10:53:25	0.57	0.03	0.54
2	1	No. 2, Heading	8,100	Exhausting	1	10:54:03 to 10:55:50	1.17	0.06	1.11
2	1	No. 2, Heading	8,100	Exhausting	2	10:56:45 to 10:58:25	0.96	0.05	0.91
2	1	No. 2, Heading	8,100	Exhausting	2	11:03:25 to 11:04:50	0.82	0.14	0.69
2	1	No. 2, Heading	8,100	Exhausting	1	11:07:40 to 11:09:10	0.49	0.11	0.37
2	1	No. 2, Heading	8,100	Exhausting	2	11:13:10 to 11:14:55	0.88	0.10	0.77
2	1	No. 2, Heading	8,100	Exhausting	2	12:00:48 to 12:01:49	0.40	0.12	0.28
2	1	No. 2, Heading	8,100	Exhausting	1	12:04:17 to 12:05:50	0.49	0.13	0.36
2	1	No. 2, Heading	8,100	Exhausting	2	12:06:12 to 12:07:39	0.99	0.09	0.90
2	1	No. 2, Heading	8,100	Exhausting	1	12:09:07 to 12:10:28	0.06	0.04	0.02
2	1	No. 2, Heading	8,100	Exhausting	2	12:10:49 to 12:12:32	0.70	0.11	0.60
2	1	No. 2, Heading	8,100	Exhausting	1	12:12:52 to 12:14:39	1.44	0.04	1.40
2	1	No. 2, Heading	8,100	Exhausting	1	12:23:40 to 12:25:15	0.56	0.05	0.50
2	1	No. 2, Heading	8,100	Exhausting	2	12:25:34 to 12:27:05	1.61	0.05	1.55
2	1	No. 2, Heading	8,100	Exhausting	2	12:29:35 to 12:31:05	0.40	0.03	0.38
2	1	No. 2, Heading	8,100	Exhausting	1	12:31:36 to 12:33:46	0.86	0.03	0.83
2	1	No. 2, Heading	8,100	Exhausting	2	12:35:22 to 12:36:15	1.17	0.03	1.14
2	1	No. 2, Heading	8,100	Exhausting	2	12:37:37 to 12:38:40	1.34	0.06	1.28
3	1	No. 1, Heading	5,100	Exhausting	2	13:00:30 to 13:01:56	0.24	0.39	0.00
3	1	No. 1, Heading	5,100	Exhausting	2	13:04:10 to 13:05:30	0.32	0.75	0.00
3	1	No. 1, Heading	5,100	Exhausting	2	13:08:15 to 13:09:30	0.63	0.39	0.24
3	1	No. 1, Heading	5,100	Exhausting	2	13:11:40 to 13:12:54	1.64	0.89	0.76
3	1	No. 1, Heading	5,100	Exhausting	2	13:15:15 to 13:16:42	1.83	0.95	0.89
3	1	No. 1, Heading	5,100	Exhausting	2	13:18:54 to 13:21:08	0.44	0.80	0.00
3	1	No. 1, Heading	5,100	Exhausting	2	13:23:20 to 13:25:05	1.52	0.47	1.05
3	1	No. 1, Heading	5,100	Exhausting	2	13:27:14 to 13:28:51	3.73	0.31	3.41
3	1	No. 1, Heading	5,100	Exhausting	2	13:32:30 to 13:33:46	0.82	0.13	0.70
3	1	No. 1, Heading	5,100	Exhausting	2	13:38:24 to 13:39:36	0.51	0.13	0.38
3	1	No. 1, Heading	5,100	Exhausting	2	13:41:53 to 13:43:15	0.64	0.12	0.52
3	1	No. 1, Heading	5,100	Exhausting	2	13:47:05 to 13:49:02	0.13	0.10	0.02
3	1	No. 1, Heading	5,100	Exhausting	2	13:51:33 to 13:53:10	0.71	0.06	0.66
3	1	No. 1, Heading	5,100	Exhausting	2	13:59:05 to 14:01:10	0.21	0.16	0.05
3	1	No. 1, Heading	5,100	Exhausting	1	14:01:55 to 14:04:15	0.33	0.12	0.20
3	1	No. 1, Heading	5,100	Exhausting	2	14:06:59 to 14:08:51	0.37	0.10	0.27
3	1	No. 1, Heading	5,100	Exhausting	1	14:09:30 to 14:11:25	1.13	0.10	1.03

Table B-2. Shuttle car dust levels when loading at the face during Cuts No. 4 through 6

Note: Shaded areas show deep-cut depth, and clear areas show regular-cut depth.
Airflows are before scrubber activation.

Cut No.	Day	Cut Sequence	Curtain Airflow (cfm)	Face Ventilation	Car No.	Time at Face	Shuttle Car Dust (mg/m³)	Intake Dust Level (mg/m³)	Adjusted Shuttle Car Dust (mg/m³)
4	2	No. 4, Heading	0	No Curtain	1	7:54:13 to 7:56:26	0.23	0.38	0.00
4	2	No. 4, Heading	0	No Curtain	2	7:57:06 to 8:00:13	0.23	0.11	0.12
4	2	No. 4, Heading	0	No Curtain	1	8:01:39 to 8:04:10	0.59	0.06	0.53
4	2	No. 4, Heading	0	No Curtain	2	8:05:10 to 8:08:05	0.14	0.01	0.13
4	2	No. 4, Heading	0	No Curtain	1	8:08:50 to 8:12:10	0.49	0.06	0.43
4	2	No. 4, Heading	0	No Curtain	2	8:14:24 to 8:16:20	0.13	0.09	0.03
4	2	No. 4, Heading	0	No Curtain	1	8:17:00 to 8:19:14	0.62	0.02	0.60
4	2	No. 4, Heading	0	No Curtain	2	8:20:00 to 8:22:13	0.44	0.02	0.42
4	2	No. 4, Heading	0	No Curtain	1	8:25:05 to 8:27:11	1.23	0.01	1.22
4	2	No. 4, Heading	0	No Curtain	1	8:28:45 to 8:30:43	1.36	0.02	1.34
4	2	No. 4, Heading	0	No Curtain	1	8:32:24 to 8:35:08	0.87	0.01	0.86
4	2	No. 4, Heading	0	No Curtain	2	8:35:50 to 8:38:22	0.24	0.03	0.20
4	2	No. 4, Heading	0	No Curtain	1	8:39:10 to 8:41:42	0.68	0.06	0.63
4	2	No. 4, Heading	0	No Curtain	2	8:44:02 to 8:47:00	0.13	0.02	0.11
4	2	No. 4, Heading	0	No Curtain	1	8:47:42 to 8:49:33	0.75	0.02	0.73
4	2	No. 4, Heading	0	No Curtain	2	8:50:11 to 8:52:19	0.32	0.03	0.29
4	2	No. 4, Heading	0	No Curtain	1	8:53:22 to 8:55:50	0.91	0.03	0.88
5	2	No. 3, Heading	9,000	Exhausting	2	10:10:03 to 10:11:15	0.06	0.47	0.00
5	2	No. 3, Heading	9,000	Exhausting	1	10:11:48 to 10:12:46	0.13	0.05	0.08
5	2	No. 3, Heading	9,000	Exhausting	2	10:13:36 to 10:14:35	0.17	0.03	0.14
5	2	No. 3, Heading	9,000	Exhausting	1	10:17:00 to 10:18:32	0.13	0.03	0.10
5	2	No. 3, Heading	9,000	Exhausting	2	10:19:09 to 10:20:18	0.15	0.03	0.12
5	2	No. 3, Heading	9,000	Exhausting	2	10:24:20 to 10:25:47	0.15	0.05	0.10
5	2	No. 3, Heading	9,000	Exhausting	1	10:26:21 to 10:29:02	0.12	0.03	0.09
5	2	No. 3, Heading	9,000	Exhausting	2	10:29:57 to 10:34:57	0.67	0.04	0.62
5	2	No. 3, Heading	9,000	Exhausting	1	10:35:38 to 10:40:56	0.70	0.06	0.63
5	2	No. 3, Heading	9,000	Exhausting	2	10:41:33 to 10:44:19	0.73	0.07	0.66
5	2	No. 3, Heading	9,000	Exhausting	1	10:45:03 to 10:51:03	0.92	0.08	0.84
5	2	No. 3, Heading	9,000	Exhausting	2	10:51:43 to 10:54:55	1.24	0.05	1.19
5	2	No. 3, Heading	9,000	Exhausting	1	10:55:44 to 10:59:14	0.65	0.07	0.59
5	2	No. 3, Heading	9,000	Exhausting	2	11:00:19 to 11:04:02	0.13	0.05	0.09
5	2	No. 3, Heading	9,000	Exhausting	1	11:05:43 to 11:07:40	0.44	0.04	0.41
5	2	No. 3, Heading	9,000	Exhausting	2	11:08:24 to 11:10:44	0.26	0.04	0.22
5	2	No. 3, Heading	9,000	Exhausting	1	11:11:29 to 11:13:08	0.55	0.04	0.51
5	2	No. 3, Heading	9,000	Exhausting	2	11:14:16 to 11:16:20	0.34	0.03	0.31
5	2	No. 3, Heading	9,000	Exhausting	1	11:17:09 to 11:20:07	0.55	0.10	0.45
5	2	No. 3, Heading	9,000	Exhausting	2	11:20:51 to 11:23:44	0.30	0.28	0.02
5	2	No. 3, Heading	9,000	Exhausting	1	11:26:16 to 11:33:17	0.41	0.71	0.00
5	2	No. 3, Heading	9,000	Exhausting	2	11:34:25 to 11:35:42	0.34	0.17	0.17
5	2	No. 3, Heading	9,000	Exhausting	1	11:36:30 to 11:38:05	0.65	0.10	0.54
5	2	No. 3, Heading	9,000	Exhausting	2	11:38:56 to 11:40:51	0.26	0.10	0.16
5	2	No. 3, Heading	9,000	Exhausting	1	11:42:14 to 11:44:46	0.38	0.07	0.30
5	2	No. 3, Heading	9,000	Exhausting	2	11:46:08 to 11:47:35	0.28	0.11	0.16
6	2	No. 2, Left	0	No Curtain	2	12:51:19 to 12:52:12	0.08	0.17	0.00
6	2	No. 2, Left	0	No Curtain	2	12:55:00 to 12:55:51	0.16	0.23	0.00
6	2	No. 2, Left	0	No Curtain	2	12:58:07 to 12:59:20	1.15	0.30	0.84
6	2	No. 2, Left	0	No Curtain	2	13:01:42 to 13:02:42	1.48	0.40	1.08
6	2	No. 2, Left	0	No Curtain	2	13:04:56 to 13:06:00	0.10	0.40	0.00
6	2	No. 2, Left	0	No Curtain	2	13:09:19 to 13:10:39	0.12	0.20	0.00
6	2	No. 2, Left	0	No Curtain	2	13:13:02 to 13:14:40	0.08	0.05	0.04
6	2	No. 2, Left	0	No Curtain	1	13:15:30 to 13:16:40	0.26	0.03	0.22
6	2	No. 2, Left	0	No Curtain	2	13:17:15 to 13:18:23	0.18	0.12	0.07
6	2	No. 2, Left	0	No Curtain	1	13:19:11 to 13:20:30	0.24	0.05	0.19
6	2	No. 2, Left	0	No Curtain	2	13:21:05 to 13:22:18	0.17	0.07	0.10
6	2	No. 2, Left	0	No Curtain	1	13:23:20 to 13:24:42	0.27	0.06	0.21
6	2	No. 2, Left	0	No Curtain	2	13:25:26 to 13:26:44	0.14	0.05	0.10
6	2	No. 2, Left	0	No Curtain	1	13:27:26 to 13:28:39	0.31	0.02	0.29
6	2	No. 2, Left	0	No Curtain	2	13:31:12 to 13:33:05	0.02	0.02	0.00
6	2	No. 2, Left	0	No Curtain	1	13:33:50 to 13:35:47	0.21	0.02	0.19
6	2	No. 2, Left	0	No Curtain	1	13:48:05 to 13:49:35	0.07	0.08	0.00
6	2	No. 2, Left	0	No Curtain	2	13:50:21 to 13:51:33	0.13	0.09	0.04

Table B-3. Miner-generated dust levels during the regular- and deep-cut depths for each cut
Note: Dust levels are adjusted for productivity.

Cut No.	Cut Sequence	Depth	Curtain Airflow (cfm)	Face Ventilation	Starting Scrubber Airflow (cfm)	Ending Scrubber Airflow (cfm)	Miner Intake Dust Level (mg/m^3)	Miner Return Dust Level (mg/m^3)	Miner Generated (mg/m^3)	Cars Per Minute	Adjusted Miner Generated (mg/m^3)
1	No. 1, heading	Regular	0	No Curtain	7,020	N/A	0.08	1.55	1.47	0.25	1.82
1	No. 1, heading	Deep	0	No Curtain	N/A	6,900	0.12	1.27	1.15	0.24	1.48
2	No. 2, heading	Regular	8,100	Exhausting	7,150	N/A	0.09	0.75	0.67	0.34	0.61
2	No. 2, heading	Deep	8,100	Exhausting	N/A	7,150	0.05	0.87	0.81	0.43	0.59
3	No. 1, heading	Regular	5,130	Exhausting	7,390	N/A	0.55	1.19	0.64	0.27	0.73
3	No. 1, heading	Deep	5,130	Exhausting	N/A	4,820	0.11	1.38	1.27	0.30	1.31
4	No. 4, heading	Regular	0	No Curtain	7,670	N/A	0.07	14.78	14.71	0.27	16.89
4	No. 4, heading	Deep	0	No Curtain	N/A	6,080	0.03	1.76	1.73	0.30	1.78
5	No. 3, heading	Regular	9,000	Exhausting	7,560	N/A	0.06	5.54	5.48	0.29	5.85
5	No. 3, heading	Deep	9,000	Exhausting	N/A	6,900	0.20	2.11	1.91	0.27	2.19
6	No. 2, left	Regular	0	No Curtain	7,940	N/A	0.25	1.51	1.26	0.33	1.18
6	No. 2, left	Deep	0	No Curtain	N/A	7,500	0.05	2.61	2.56	0.45	1.76

Table B-4. Miner operator dust levels during the regular- and deep-cut depths for each cut
Note: Dust levels are adjusted for productivity.

Cut No.	Cut Sequence	Depth	Curtain Airflow (cfm)	Face Ventilation	Starting Scrubber Airflow (cfm)	Ending Scrubber Airflow (cfm)	Miner Operator Exposure (mg/m^3)	Cars Per Minute	Adj. Miner Operator Exposure (mg/m^3)
1	No. 1, heading	Regular	0	No Curtain	7,020	N/A	0.16	0.25	0.20
1	No. 1, heading	Deep	0	No Curtain	N/A	6,900	0.16	0.24	0.20
2	No. 2, heading	Regular	8,100	Exhausting	7,150	N/A	0.27	0.34	0.25
2	No. 2, heading	Deep	8,100	Exhausting	N/A	7,150	0.48	0.43	0.35
3	No. 1, heading	Regular	5,130	Exhausting	7,390	N/A	0.97	0.27	1.11
3	No. 1, heading	Deep	5,130	Exhausting	N/A	4,820	0.87	0.30	0.90
4	No. 4, heading	Regular	0	No Curtain	7,670	N/A	0.99	0.27	1.14
4	No. 4, heading	Deep	0	No Curtain	N/A	6,080	0.20	0.30	0.21
5	No. 3, heading	Regular	9,000	Exhausting	7,560	N/A	0.54	0.29	0.58
5	No. 3, heading	Deep	9,000	Exhausting	N/A	6,900	0.61	0.27	0.70
6	No. 2, left	Regular	0	No Curtain	7,940	N/A	0.74	0.33	0.69
6	No. 2, left	Deep	0	No Curtain	N/A	7,500	0.50	0.45	0.34

Table B-5. Bolter operator exposure levels during the regular- and deep-cut depths for each bolted room

Room No.	Entry	Depth	Position With Respect To Miner	Curtain Airflow (cfm)	Face Ventilation	Bolter Suction Left/Right (in Hg)	Left-Side Bolter Dust Levels (mg/m^3)	Right-Side Bolter Dust Levels (mg/m^3)	Upwind Bolter Dust Levels (mg/m^3)	Downwind Bolter Dust Levels (mg/m^3)	Bolter Generated Dust (mg/m^3)
1	No. 6, right	Regular	Downwind	2,920	Exhausting	12/15	0.00	0.21	Invalid	Invalid	Invalid
1	No. 6, right	Deep	Downwind	2,920	Exhausting	12/15	0.24	0.46	Invalid	Invalid	Invalid
2	No. 6, right	Regular	Upwind	1,840	Exhausting	15/13	0.00	0.29	0.08	0.27	0.19
2	No. 6, right	Deep	Upwind	1,840	Exhausting	15/13	0.17	0.19	0.05	0.07	0.02
3	No. 5, heading	Regular	Upwind	4,680	Exhausting	15/13	0.73	0.35	0.07	0.17	0.10
3	No. 5, heading	Deep	Upwind	4,680	Exhausting	15/13	Battery	0.59	0.02	0.13	0.11
4	No. 6, heading	Regular	Downwind	3,600	Exhausting	15/17	0.33	0.68	Invalid	Invalid	Invalid
4	No. 6, heading	Deep	Downwind	3,600	Exhausting	15/17	0.39	0.81	Invalid	Invalid	Invalid
5	No. 7, heading	Regular	Downwind	3,150	Exhausting	15/16	0.68	1.22	Invalid	Invalid	Invalid
5	No. 7, heading	Deep	Downwind	3,150	Exhausting	15/16	1.95	2.08	Invalid	Invalid	Invalid

Appendix C: MINE C CASE STUDY

Mine-specific Information

Mine C used a Joy 12CM continuous miner on this section. A total of 50 sprays were operated—33 dust suppression sprays were operated at a pressure of not less than 100 psi. Additionally, 14 cooler sprays, along with 2 scrubber sprays and 1 fogger spray, were operated at 50 psi. The minimum numbers of sprays required by the mine ventilation plan to be operational were 29 dust suppression, 12 cooler, 2 scrubber, and 1 fogger sprays. The miner spray configuration is shown in Figure C-1.

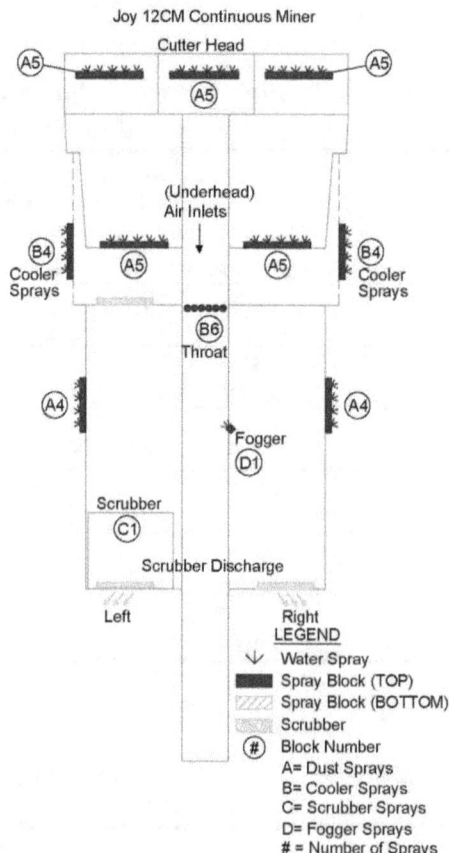

Figure C-1. Mine C continuous miner spray configuration.

Figure C-2 shows a plan view of the continuous miner section surveyed for this study. The section used a sweeping ventilation configuration, with Entry No. 5 serving as the intake and Entry No. 1 serving as the return. All other entries were neutral, including the beltway, which was located in Entry No. 3. Mining height varied between 106 and 120-in, with entry width averaging 20 ft. The mine used a combination of exhaust tubing and curtain to ventilate the active faces, which was required to be no more than 40 ft from the unbolted face. The mine ventilation plan required airflows at the face when cutting coal to be a minimum of 8000 cfm with the scrubber off and 12000 cfm with the scrubber on. When cutting rock, the minimum airflows at the face are set at 11000 cfm with the scrubber off and 13000 cfm with the scrubber on. Tubing was used on Days 1 and 2 of the study, and curtain was used on Day 3. Pillar dimensions were approximately 110 x 70 ft. The deep-cut depths were 40 ft. The cut sequence for Mine C was as follows:

   (1) 20-ft sump cut, left side
   (2) 20-ft slab cut, right side
   (3) 20-ft sump cut, right side
   (4) 20-ft slab cut, left side

The Mine C dust survey was conducted on the June 9, 10, and 11, 2009.

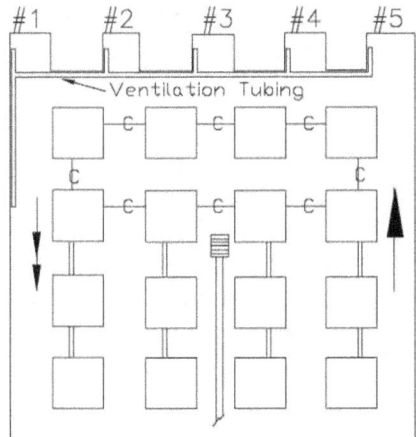

Figure C-2. Plan view of the mine C continuous miner development section.

Shuttle Car Data Analysis

Tables C-1, C-2, and C-3 show the shuttle car data collected during the study. Mine C used a Joy 12CM miner on the section with two 12-ton Joy shuttle cars. Shuttle Car Cab No. 2 had a left-side (off standard) operator's cab, while Car No. 1 had a right-side (standard) cab. Two unique types of deep cuts were evaluated for this study:

   (1) Continuous miner driving a heading with exhausting face ventilation
   (2) Continuous miner turning a right crosscut with exhausting face ventilation

These two types of deep-cuts represented the only types of cuts used on this section as all crosscuts were turned toward intake air.

Throughout the study, dust levels attributable to face operations, as measured in the shuttle car cabs, were very low, averaging from 0.05 to 0.48 mg/m^3. When driving a heading using exhausting face ventilation, dust levels in the standard cab attributable to face operations were not significantly (85% CIs) different when comparing the regular-cut depth average of 0.06

mg/m³ to the deep-cut depth average of 0.12 mg/m³. For the off-standard cab, dust levels were significantly (95% CIs) higher for deep-cut sequence, averaging 0.07 mg/m³ during regular-cut depths and 0.31 mg/m³ during deep-cut depths. A similar phenomenon was observed during development of the right cross-cuts. When driving the cross-cuts, dust levels from face operations in the standard cab were not influenced by depth of cut, averaging 0.05 mg/m³ during the regular-cut depths and 0.07 mg/m³ during the deep-cut depths. However, dust levels in the off-standard cab when developing the cross-cuts were, again, significantly (90% CIs) higher during the deep-cut sequence, averaging 0.11 mg/m³ during the regular-cut depth and 0.48 mg/m³ during the deep-cut depth. In general, dust levels in the off-standard cab attributable to face operations during loading were 0.2 to 0.4 mg/m³ higher during the deep-cut sequence. This may have resulted because the off-standard cab was on the same side of the entry as the exhaust tubing/curtain. As the cut advanced, the cab got closer to the mouth of the tubing/curtain and may have experienced some dust rollback.

The daily average dust levels for the shuttle car cabs were very low throughout the study, ranging from 0.13 to 0.39 mg/m³, indicating that use of deep-cutting techniques did not hinder the mine's ability to limit the exposure levels of the shuttle car operators.

Miner-generated Dust

Table C-4 shows miner-generated dust levels during the regular- and deep-cut sequences for each room development. Miner-generated dust levels could not be determined during Day 1 (Cuts No. 1 and 2) of the study because an automatic rock dusting device was attached to the return end of the exhaust tubing, which resulted in contamination of the downwind samples. When tubing was used to ventilate the active faces, which occurred on Day 1 (Cuts No. 1 and 2) and Day 2 (Cuts No. 3 through 5) of the study, the continuous miner and bolting machine shared a common return; however, the miner return samples were still deemed to be valid on Day 2 of the study due to the negligible amount of dust generated by the bolting machine. The bolting machine was found to generate less than 0.10 mg/m³ during operation on Day 3 of the study. The last column of Table C-4 normalizes the data to allow a comparison of miner-generated dust during the regular- and deep-cut depths by adjusting dust levels based on productivity. The average number of cars loaded per minute for the section was 0.40.

Due to the invalidity of miner return data collected on Day 1, only one right crosscut development was analyzed for miner dust generation—Cut No. 6 on Day 3. For this particular cut, dust levels were higher during the deep-cut depth, averaging 1.32 mg/m³ during the regular-cut depth and 4.62 mg/m³ during the deep-cut depth. This particular cut was ventilated with exhaust curtain so it is possible that airflow levels fluctuated during the cut. However, for safety reasons related to shuttle car traffic, researchers were unable to periodically monitor airflow levels behind the curtain during production. When all data were grouped together, miner-generated dust levels were not significantly (85% CIs) different when comparing the average of 1.77 mg/m³ during the regular-cut depth to the average of 2.19 mg/m³ during the deep-cut depth.

Miner Operator Data Analysis

Table C-5 shows the miner operator exposure levels as measured with a PDM device during each cut of the study for both the regular- and deep-cut depths. These exposures are net of intake dust levels and represent exposures resulting from face operations. The last column of Table C-5 shows the values normalized for productivity. The operator's exposure level was very low for both the regular-cut and deep-cut sequences. When driving a heading, the operator's exposure level averaged 0.33 mg/m^3 during the regular-cut depth and 0.55 mg/m^3 during the deep-cut depth. When developing the right crosscut, the miner operator's dust exposure level averaged 0.57 mg/m^3 during the regular-cut depth and 0.92 mg/m^3 during the deep-cut depth. Neither unique cut produced a statistically significant (85% CIs) variation in dust levels when comparing regular- to deep-cut depth. When all data were grouped together, miner operator dust exposure levels during production were not significantly (85% CIs) different when comparing the average dust level of 0.42 mg/m^3 during the regular-cut depth to 0.69 mg/m^3 during the deep-cut depth.

The daily average dust exposure levels for the miner operators were very low throughout the study, ranging from 0.34 to 0.57 mg/m^3, indicating that use of deep-cutting techniques did not hinder the mine's ability to limit the exposure levels of the miner operators.

Bolter Operations Data Analysis

Table C-6 shows bolter operator dust exposure levels during the regular- and deep-cut sequences for each room that was bolted. Exposures were determined from PDM devices worn by the operators. The right-side operator's PDM malfunctioned on Day 1 (Rooms No. 1 and 2) of the study and experienced battery failure during bolting operations in the last room surveyed for this study, resulting in lost data as indicated in the table. Bolter intake and return data are also displayed in Table C-6. Bolter return data for Days 1 and 2 (Rooms No. 1 through 5) were invalid due to contamination from rock dust and miner-generated dust on those days. On Day 3, exhausting curtain was used instead of tubing so the bolter had an independent return, allowing isolation of machine-generated dust for Rooms No. 6 and 7.

After adjusting for intake dust levels, the bolter operators' exposure levels to dust from machine operations averaged less than 0.10 mg/m^3 during the 3-day study. Likewise, bolter-generated dust as measured in the return and net of intake dust levels averaged less than 0.10 mg/m^3. Mine C used a Fletcher DDR-13-A bolter equipped with wet drilling technology, which appears to be a very effective method for limiting operator exposure to dust. One concern associated with a dry dust collection system is that its filtration system may become over-loaded with dust during bolting in deep-cut rooms, resulting in less dust capture and higher operator exposure levels as the bolter is advanced. Use of wet drilling eliminates this concern.

The daily average dust exposure levels for the bolter operators were very low throughout the study, ranging from 0.11 to 0.55 mg/m^3, indicating that use of deep-cutting techniques did not hinder the mine's ability to limit the exposure levels of the bolter operators.

Dust-control Monitoring

Face ventilation at Mine C, using tubing and curtain, was well maintained throughout the study. On the miner faces, the tubing was initially set about four roof bolts back from the face and was advanced at least once during the cut to the last row of bolts. The miner operator properly maintained his position parallel to or slightly outby the mouth of the tubing. The exhaust curtain was maintained in a similar fashion. Mine C has a variance permitting the scrubber to be exhausted on the right side of the machine and toward the face, which allows the mine to maintain its curtain closer to the face without interference from the scrubber discharge. This technique appeared to be highly effective for dust control at this particular mine. Face ventilation airflows are shown in Tables C-4, C-5, and C-6. Average airflow on the miner faces was 17,400 cfm, ranging from 14,300 to 21,500 cfm. Average airflow on the bolting faces was 3,900 when using tubing and 10,300 when using curtain. Primary intake airflow was approximately 27,000 cfm during our study.

As mentioned earlier, one of the concerns was that the scrubber would become clogged with particles as the mining cycle progressed through the deep cut, diminishing its effectiveness. Scrubber airflows, measured at the start and end of each cut, are shown in Tables C-4 and C-5. The scrubber was not used during the first half of Cut No. 1 because the mine has a variance to de-activate the scrubber until it clears the right-side rib during crosscut development. Diminishing performance was minimal at Mine C, as the average scrubber airflow dropped only 6% from start to finish during the cuts. The scrubber screen was tapped and back-flushed before each cut, which is a practice that should be continued to assure proper scrubber function.

Summary

1. When driving a heading using exhausting face ventilation, dust levels in the standard-cab shuttle car attributable to face operations were not significantly (85% CIs) different when comparing the regular-cut depth to the deep-cut depth, averaging 0.06 mg/m^3 during the regular-cut depth and 0.12 mg/m^3 during the deep-cut depth.

2. When driving a heading using exhausting face ventilation, dust from face operations in the off-standard cab were significantly (95% CIs) higher for the deep-cut sequence, averaging 0.07 mg/m^3 during the regular-cut depth and 0.31 mg/m^3 during the deep-cut depth.

3. When driving the right crosscuts with exhausting face ventilation, dust levels from face operations in the standard cab were not influenced by depth of cut, averaging 0.05 mg/m^3 during the regular-cut depth and 0.07 mg/m^3 during the deep-cut depth.

4. When driving the right crosscuts with exhausting face ventilation, dust from face operations in the off-standard cab were significantly (90% CIs) higher during the deep-cut sequence, averaging 0.11 mg/m^3 during the regular-cut depth and 0.48 mg/m^3 during the deep-cut depth.

5. In general, dust levels in the off-standard cab attributable to face operations during loading were 0.20 to 0.40 mg/m^3 higher during the deep-cut sequence. This may have

resulted because the off-standard cab was on the same side of the entry as the exhaust tubing/curtain. As the cut advanced, the cab got closer to the mouth of the tubing/curtain and may have experienced some dust rollback.

6. The daily average dust levels for the shuttle car cabs were very low throughout the study, ranging from 0.13 to 0.39 mg/m^3, indicating that use of deep-cutting techniques did not hinder the mine's ability to limit the exposure levels of the shuttle car operators.

7. Miner-generated dust levels were not significantly (85% CIs) different when comparing the regular- to deep-cut depth, averaging 1.77 during the regular-cut depth and 2.19 mg/m^3 during the deep-cut depth.

8. Miner operator dust exposure levels were not significantly (85% CIs) different when comparing the regular- to deep-cut depth, averaging 0.42 during the regular-cut depth and 0.69 mg/m^3 during the deep-cut depth.

9. The daily average dust exposure levels for the miner operators were very low throughout the study, ranging from 0.34 to 0.57 mg/m^3, indicating that use of deep-cutting techniques did not hinder the mine's ability to limit the exposure levels of the miner operators.

10. After adjusting for intake dust levels, the bolter operators' exposure levels to dust from bolting operations averaged less than 0.10 mg/m^3 during the 3-day study. Likewise, bolter-generated dust as measured in the return and net of intake dust levels averaged less than 0.10 mg/m^3.

11. The daily average dust exposure levels for the bolter operators were very low throughout the study, ranging from 0.11 to 0.55 mg/m^3, indicating that use of deep-cutting techniques did not hinder the mine's ability to limit the exposure levels of the bolter operators.

Table C-1. Shuttle car dust levels when loading at the face during Day 1 of the study

Note: The bolded data rows are excluded from the reported average dust levels because they occurred after the miner cut through to the adjacent entry. Shaded areas show deep-cut depth, and clear areas show regular-cut depth. Airflows are before scrubber activation.

Cut No.	Day	Cut Sequence	Curtain or Tubing Airflow	Face Ventilation	Car No.	Time at Face	Shuttle Car Dust (mg/m^3)	Intake Dust Level (mg/m^3)	Adjusted Shuttle Car Dust (mg/m^3)
1	1	No. 1, Right	16,635	Exhaust Tubing	2	10:10:16 to 10:11:01	0.14	0.18	0.00
1	1	No. 1, Right	16,635	Exhaust Tubing	1	10:14:00 to 10:14:45	0.14	0.18	0.00
1	1	No. 1, Right	16,635	Exhaust Tubing	2	10:17:02 to 10:17:40	0.16	0.10	0.06
1	1	No. 1, Right	16,635	Exhaust Tubing	1	10:22:39 to 10:23:36	0.14	0.07	0.06
1	1	No. 1, Right	16,635	Exhaust Tubing	2	10:25:43 to 10:26:26	0.12	0.09	0.03
1	1	No. 1, Right	16,635	Exhaust Tubing	1	10:28:58 to 10:30:15	0.13	0.09	0.04
1	1	No. 1, Right	16,635	Exhaust Tubing	2	10:32:47 to 10:34:03	0.07	0.05	0.02
1	1	No. 1, Right	16,635	Exhaust Tubing	1	10:36:31 to 10:37:59	0.14	0.15	0.00
1	1	No. 1, Right	16,635	Exhaust Tubing	2	10:40:53 to 10:42:12	0.14	0.04	0.10
1	1	No. 1, Right	16,635	Exhaust Tubing	1	10:44:39 to 10:45:17	0.23	0.27	0.00
1	1	No. 1, Right	16,635	Exhaust Tubing	2	10:47:21 to 10:48:10	1.43	0.10	1.33
1	1	No. 1, Right	16,635	Exhaust Tubing	1	10:50:58 to 10:51:33	0.31	0.77	0.00
1	1	No. 1, Right	16,635	Exhaust Tubing	2	10:57:01 to 10:57:50	0.92	0.15	0.76
1	1	No. 1, Right	16,635	Exhaust Tubing	1	11:01:08 to 11:01:58	0.15	0.12	0.03
1	1	No. 1, Right	16,635	Exhaust Tubing	2	11:04:13 to 11:05:00	0.66	0.04	0.62
1	1	No. 1, Right	16,635	Exhaust Tubing	1	11:08:31 to 11:10:06	0.15	0.21	0.00
1	1	No. 1, Right	16,635	Exhaust Tubing	2	11:12:41 to 11:13:30	1.57	0.06	1.51
1	1	No. 1, Right	16,635	Exhaust Tubing	1	11:16:08 to 11:16:51	0.30	0.11	0.20
1	1	No. 1, Right	16,635	Exhaust Tubing	2	11:19:34 to 11:20:07	0.24	0.11	0.12
1	1	No. 1, Right	16,635	Exhaust Tubing	1	11:22:41 to 11:23:28	0.19	0.18	0.01
2	1	No. 2, Right	17,077	Exhaust Tubing	2	12:43:43 to 12:44:50	0.36	0.07	0.29
2	1	No. 2, Right	17,077	Exhaust Tubing	1	12:47:55 to 12:48:29	0.19	0.17	0.02
2	1	No. 2, Right	17,077	Exhaust Tubing	2	12:50:35 to 12:51:00	0.33	0.10	0.23
2	1	No. 2, Right	17,077	Exhaust Tubing	1	12:53:15 to 12:54:00	0.27	0.15	0.12
2	1	No. 2, Right	17,077	Exhaust Tubing	2	12:55:43 to 12:56:18	0.24	0.18	0.07
2	1	No. 2, Right	17,077	Exhaust Tubing	1	12:58:18 to 12:59:01	0.14	0.04	0.10
2	1	No. 2, Right	17,077	Exhaust Tubing	2	13:00:52 to 13:01:30	0.42	0.05	0.37
2	1	No. 2, Right	17,077	Exhaust Tubing	1	13:05:24 to 13:06:09	0.17	0.04	0.13
2	1	No. 2, Right	17,077	Exhaust Tubing	2	13:08:20 to 13:09:26	0.20	0.25	0.00
2	1	No. 2, Right	17,077	Exhaust Tubing	1	13:14:59 to 13:15:32	0.18	0.10	0.08
2	1	No. 2, Right	17,077	Exhaust Tubing	2	13:17:32 to 13:18:07	0.25	0.10	0.15
2	1	No. 2, Right	17,077	Exhaust Tubing	1	13:20:07 to 13:20:48	0.16	0.05	0.12
2	1	No. 2, Right	17,077	Exhaust Tubing	2	13:22:47 to 13:23:21	0.14	0.07	0.07
2	1	No. 2, Right	17,077	Exhaust Tubing	1	13:25:54 to 13:27:33	0.14	0.02	0.12
2	1	No. 2, Right	17,077	Exhaust Tubing	2	13:29:36 to 13:30:09	0.14	0.04	0.10
2	1	No. 2, Right	17,077	Exhaust Tubing	1	13:32:21 to 13:33:35	0.14	0.03	0.11
2	**1**	**No. 2, Right**	**17,077**	**Exhaust Tubing**	**2**	**13:35:41 to 13:37:46**	**0.24**	**0.94**	**0.00**
2	**1**	**No. 2, Right**	**17,077**	**Exhaust Tubing**	**1**	**13:39:59 to 13:40:41**	**0.22**	**0.46**	**0.00**

Table C-2. Shuttle car dust levels when loading at the face during Day 2 of the study
Note: Shaded areas are deep cuts, clear areas are regular cuts. Airflows are before scrubber activation.

Cut No.	Day	Cut Sequence	Curtain or Tubing Airflow	Face Ventilation	Car No.	Time at Face	Shuttle Car Dust (mg/m^3)	Intake Dust Level (mg/m^3)	Adjusted Shuttle Car Dust (mg/m^3)
3	2	No. 4, Heading	21,522	Exhaust Tubing	2	9:16:24 to 9:17:00	0.08	0.04	0.04
3	2	No. 4, Heading	21,522	Exhaust Tubing	1	9:19:54 to 9:20:21	0.11	0.02	0.09
3	2	No. 4, Heading	21,522	Exhaust Tubing	2	9:21:04 to 9:21:36	0.07	0.03	0.05
3	2	No. 4, Heading	21,522	Exhaust Tubing	1	9:23:16 to 9:23:50	0.10	0.02	0.08
3	2	No. 4, Heading	21,522	Exhaust Tubing	2	9:25:07 to 9:25:40	0.13	0.05	0.08
3	2	No. 4, Heading	21,522	Exhaust Tubing	1	9:27:10 to 9:28:27	0.11	0.05	0.06
3	2	No. 4, Heading	21,522	Exhaust Tubing	2	9:28:58 to 9:29:42	0.15	0.07	0.08
3	2	No. 4, Heading	21,522	Exhaust Tubing	1	9:32:00 to 9:33:31	0.10	0.08	0.02
3	2	No. 4, Heading	21,522	Exhaust Tubing	2	9:34:00 to 9:34:50	0.12	0.12	0.00
3	2	No. 4, Heading	21,522	Exhaust Tubing	1	9:36:20 to 9:37:03	0.10	0.08	0.02
3	2	No. 4, Heading	21,522	Exhaust Tubing	2	9:38:30 to 9:39:36	0.13	0.05	0.07
3	2	No. 4, Heading	21,522	Exhaust Tubing	1	9:41:25 to 9:42:35	0.17	0.02	0.15
3	2	No. 4, Heading	21,522	Exhaust Tubing	2	9:43:32 to 9:44:10	0.11	0.02	0.08
3	2	No. 4, Heading	21,522	Exhaust Tubing	1	9:45:35 to 9:46:24	0.08	0.03	0.05
3	2	No. 4, Heading	21,522	Exhaust Tubing	2	9:47:14 to 9:48:04	0.12	0.04	0.07
3	2	No. 4, Heading	21,522	Exhaust Tubing	1	9:49:26 to 9:50:26	0.10	0.05	0.05
3	2	No. 4, Heading	21,522	Exhaust Tubing	2	9:51:20 to 9:52:09	2.93	0.03	2.89
3	2	No. 4, Heading	21,522	Exhaust Tubing	1	9:54:12 to 9:55:55	0.06	0.07	0.00
3	2	No. 4, Heading	21,522	Exhaust Tubing	2	9:58:02 to 9:59:20	0.15	0.06	0.09
3	2	No. 4, Heading	21,522	Exhaust Tubing	1	10:00:13 to 10:01:14	0.14	0.07	0.07
3	2	No. 4, Heading	21,522	Exhaust Tubing	2	10:02:33 to 10:03:09	0.37	0.11	0.26
4	2	No. 5, Heading	16,743	Exhaust Tubing	2	10:50:32 to 10:51:50	0.08	0.06	0.02
4	2	No. 5, Heading	16,743	Exhaust Tubing	1	10:54:51 to 10:55:20	0.13	0.07	0.05
4	2	No. 5, Heading	16,743	Exhaust Tubing	1	10:55:56 to 10:56:56	0.09	0.07	0.02
4	2	No. 5, Heading	16,743	Exhaust Tubing	2	10:58:32 to 10:58:58	0.21	0.08	0.13
4	2	No. 5, Heading	16,743	Exhaust Tubing	1	11:00:16 to 11:00:50	0.12	0.06	0.06
4	2	No. 5, Heading	16,743	Exhaust Tubing	2	11:02:24 to 11:03:00	0.07	0.06	0.01
4	2	No. 5, Heading	16,743	Exhaust Tubing	1	11:03:58 to 11:04:34	0.10	0.05	0.06
4	2	No. 5, Heading	16,743	Exhaust Tubing	2	11:06:14 to 11:07:36	0.09	0.09	0.01
4	2	No. 5, Heading	16,743	Exhaust Tubing	1	11:08:15 to 11:09:02	0.39	0.10	0.29
4	2	No. 5, Heading	16,743	Exhaust Tubing	2	11:10:42 to 11:11:17	0.14	0.08	0.05
4	2	No. 5, Heading	16,743	Exhaust Tubing	1	11:11:58 to 11:12:48	0.18	0.14	0.04
4	2	No. 5, Heading	16,743	Exhaust Tubing	2	11:14:24 to 11:15:01	0.25	0.12	0.13
4	2	No. 5, Heading	16,743	Exhaust Tubing	1	11:17:04 to 11:18:23	0.08	0.07	0.01
4	2	No. 5, Heading	16,743	Exhaust Tubing	2	11:18:49 to 11:19:25	0.44	0.07	0.37
4	2	No. 5, Heading	16,743	Exhaust Tubing	1	11:21:25 to 11:21:58	0.17	0.05	0.12
4	2	No. 5, Heading	16,743	Exhaust Tubing	2	11:22:42 to 11:23:16	0.43	0.08	0.36
4	2	No. 5, Heading	16,743	Exhaust Tubing	1	11:24:55 to 11:25:32	0.21	0.11	0.10
4	2	No. 5, Heading	16,743	Exhaust Tubing	2	11:26:45 to 11:27:24	0.33	0.09	0.24
4	2	No. 5, Heading	16,743	Exhaust Tubing	1	11:28:44 to 11:31:00	0.09	0.05	0.04
4	2	No. 5, Heading	16,743	Exhaust Tubing	2	11:31:27 to 11:32:12	0.16	0.07	0.09
4	2	No. 5, Heading	16,743	Exhaust Tubing	1	11:33:49 to 11:34:30	0.18	0.03	0.15
4	2	No. 5, Heading	16,743	Exhaust Tubing	2	11:35:23 to 11:36:10	0.19	0.05	0.14
5	2	No. 4, Heading	18,325	Exhaust Tubing	2	12:57:35 to 12:58:13	0.18	0.19	0.00
5	2	No. 4, Heading	18,325	Exhaust Tubing	1	13:00:04 to 13:01:10	0.12	0.11	0.01
5	2	No. 4, Heading	18,325	Exhaust Tubing	2	13:01:44 to 13:02:41	0.17	0.07	0.10
5	2	No. 4, Heading	18,325	Exhaust Tubing	1	13:03:51 to 13:04:21	0.12	0.16	0.00
5	2	No. 4, Heading	18,325	Exhaust Tubing	2	13:05:36 to 13:06:12	0.08	0.07	0.02
5	2	No. 4, Heading	18,325	Exhaust Tubing	1	13:07:15 to 13:07:49	0.10	0.09	0.01
5	2	No. 4, Heading	18,325	Exhaust Tubing	2	13:09:10 to 13:09:40	0.22	0.08	0.15
5	2	No. 4, Heading	18,325	Exhaust Tubing	1	13:10:39 to 13:12:15	0.07	0.10	0.00
5	2	No. 4, Heading	18,325	Exhaust Tubing	2	13:27:57 to 13:28:32	0.18	0.09	0.09
5	2	No. 4, Heading	18,325	Exhaust Tubing	1	13:29:20 to 13:30:29	0.06	0.04	0.02
5	2	No. 4, Heading	18,325	Exhaust Tubing	1	13:31:36 to 13:32:23	0.14	0.08	0.06
5	2	No. 4, Heading	18,325	Exhaust Tubing	1	13:36:08 to 13:36:48	0.09	0.15	0.00
5	2	No. 4, Heading	18,325	Exhaust Tubing	2	13:37:37 to 13:38:26	0.21	0.11	0.11
5	2	No. 4, Heading	18,325	Exhaust Tubing	1	13:39:57 to 13:40:29	0.68	0.17	0.52
5	2	No. 4, Heading	18,325	Exhaust Tubing	2	13:41:28 to 13:42:17	1.02	0.06	0.96
5	2	No. 4, Heading	18,325	Exhaust Tubing	1	13:46:25 to 13:47:15	0.04	0.04	0.01
5	2	No. 4, Heading	18,325	Exhaust Tubing	2	13:48:00 to 13:48:49	0.45	0.03	0.42
5	2	No. 4, Heading	18,325	Exhaust Tubing	1	13:50:18 to 13:52:08	0.06	0.03	0.03
5	2	No. 4, Heading	18,325	Exhaust Tubing	2	13:52:50 to 13:53:58	0.05	0.02	0.03
5	2	No. 4, Heading	18,325	Exhaust Tubing	1	13:55:05 to 13:55:50	0.23	0.05	0.18
5	2	No. 4, Heading	18,325	Exhaust Tubing	2	13:57:01 to 13:57:38	0.21	0.06	0.15

Table C-3. Shuttle car dust levels when loading at the face during Day 3 of the study

Note: Shaded areas are deep cuts, clear areas are regular cuts. Airflows are before scrubber activation. Bolded text indicates when continuous miner broke through crosscut.

Cut No.	Day	Cut Sequence	Curtain or Tubing Airflow	Face Ventilation	Car No.	Time at Face	Shuttle Car Dust (mg/m³)	Intake Dust Level (mg/m³)	Adjusted Shuttle Car Dust (mg/m³)
6	3	No. 4, Right	14,336	Exhaust Curtain	1	8:39:53 to 8:41:00	0.17	0.24	0.00
6	3	No. 4, Right	14,336	Exhaust Curtain	2	8:43:55 to 8:44:45	0.09	0.09	0.01
6	3	No. 4, Right	14,336	Exhaust Curtain	1	8:46:13 to 8:46:44	0.09	0.05	0.05
6	3	No. 4, Right	14,336	Exhaust Curtain	2	8:48:05 to 8:48:35	0.18	0.04	0.14
6	3	No. 4, Right	14,336	Exhaust Curtain	1	8:50:33 to 8:51:04	0.07	0.03	0.04
6	3	No. 4, Right	14,336	Exhaust Curtain	2	8:52:50 to 8:53:18	0.14	0.02	0.12
6	3	No. 4, Right	14,336	Exhaust Curtain	1	8:55:25 to 8:56:04	0.10	0.06	0.04
6	3	No. 4, Right	14,336	Exhaust Curtain	2	8:57:44 to 8:58:17	0.17	0.01	0.15
6	3	No. 4, Right	14,336	Exhaust Curtain	1	9:00:52 to 9:01:21	0.07	0.00	0.06
6	3	No. 4, Right	14,336	Exhaust Curtain	2	9:03:38 to 9:04:57	0.10	0.00	0.10
6	3	No. 4, Right	14,336	Exhaust Curtain	1	9:06:33 to 9:07:05	0.05	0.00	0.05
6	3	No. 4, Right	14,336	Exhaust Curtain	2	9:12:57 to 9:13:31	0.19	0.00	0.18
6	3	No. 4, Right	14,336	Exhaust Curtain	1	9:16:51 to 9:17:35	0.07	0.01	0.06
6	3	No. 4, Right	14,336	Exhaust Curtain	2	9:19:15 to 9:19:44	0.23	0.03	0.20
6	3	No. 4, Right	14,336	Exhaust Curtain	1	9:22:18 to 9:23:35	0.05	0.01	0.04
6	3	No. 4, Right	14,336	Exhaust Curtain	2	9:25:22 to 9:25:56	0.25	0.00	0.25
6	3	No. 4, Right	14,336	Exhaust Curtain	1	9:28:17 to 9:29:26	0.08	0.02	0.07
6	**3**	**No. 4, Right**	**14,336**	**Exhaust Curtain**	**2**	**9:32:53 to 9:34:33**	**1.60**	**1.31**	**0.29**
6	**3**	**No. 4, Right**	**14,336**	**Exhaust Curtain**	**1**	**9:41:15 to 9:41:53**	**0.90**	**0.14**	**0.75**
6	**3**	**No. 4, Right**	**14,336**	**Exhaust Curtain**	**2**	**9:58:08 to 9:59:12**	**0.48**	**0.04**	**0.43**
7	3	No. 3, Heading	17,261	Exhaust Curtain	1	10:30:42 to 10:31:39	0.39	0.43	0.00
7	3	No. 3, Heading	17,261	Exhaust Curtain	2	10:33:02 to 10:33:31	0.19	0.25	0.00
7	3	No. 3, Heading	17,261	Exhaust Curtain	1	10:35:04 to 10:35:48	0.13	0.12	0.01
7	3	No. 3, Heading	17,261	Exhaust Curtain	2	10:45:05 to 10:45:35	0.08	0.07	0.01
7	3	No. 3, Heading	17,261	Exhaust Curtain	1	10:46:39 to 10:47:27	0.12	0.07	0.05
7	3	No. 3, Heading	17,261	Exhaust Curtain	2	10:48:30 to 10:49:08	0.19	0.08	0.11
7	3	No. 3, Heading	17,261	Exhaust Curtain	1	10:50:48 to 10:51:36	0.12	0.03	0.09
7	3	No. 3, Heading	17,261	Exhaust Curtain	2	10:53:20 to 10:54:05	0.30	0.03	0.26
7	3	No. 3, Heading	17,261	Exhaust Curtain	1	10:55:41 to 10:57:25	0.22	0.15	0.07
7	3	No. 3, Heading	17,261	Exhaust Curtain	2	11:30:08 to 11:30:28	0.13	0.13	0.00
7	3	No. 3, Heading	17,261	Exhaust Curtain	1	11:31:40 to 11:33:03	0.13	0.05	0.08
7	3	No. 3, Heading	17,261	Exhaust Curtain	2	11:34:19 to 11:35:04	0.23	0.10	0.13
7	3	No. 3, Heading	17,261	Exhaust Curtain	1	11:36:29 to 11:38:54	0.13	0.11	0.03
7	3	No. 3, Heading	17,261	Exhaust Curtain	2	11:40:05 to 11:40:37	0.21	0.04	0.16
7	3	No. 3, Heading	17,261	Exhaust Curtain	1	11:42:12 to 11:42:52	0.44	0.09	0.35
7	3	No. 3, Heading	17,261	Exhaust Curtain	2	11:44:21 to 11:45:01	0.18	0.06	0.12
7	3	No. 3, Heading	17,261	Exhaust Curtain	1	11:46:23 to 11:47:00	0.27	0.02	0.25
7	3	No. 3, Heading	17,261	Exhaust Curtain	2	11:48:12 to 11:48:52	0.88	0.03	0.86
7	3	No. 3, Heading	17,261	Exhaust Curtain	1	11:50:22 to 11:50:57	0.57	0.13	0.44
7	3	No. 3, Heading	17,261	Exhaust Curtain	2	11:52:40 to 11:54:13	0.19	0.07	0.13
7	3	No. 3, Heading	17,261	Exhaust Curtain	2	12:00:32 to 12:01:03	0.17	0.00	0.16
7	3	No. 3, Heading	17,261	Exhaust Curtain	2	12:03:50 to 12:04:33	0.25	0.00	0.24
7	3	No. 3, Heading	17,261	Exhaust Curtain	1	12:05:45 to 12:06:28	0.37	1.27	0.00
8	3	No. 2, Heading	17,177	Exhaust Curtain	1	13:28:50 to 13:29:58	0.17	0.07	0.10
8	3	No. 2, Heading	17,177	Exhaust Curtain	2	13:31:25 to 13:32:15	0.23	0.08	0.15
8	3	No. 2, Heading	17,177	Exhaust Curtain	1	13:33:52 to 13:34:59	0.16	0.04	0.12
8	3	No. 2, Heading	17,177	Exhaust Curtain	2	13:36:28 to 13:37:00	0.13	0.49	0.00
8	3	No. 2, Heading	17,177	Exhaust Curtain	1	13:38:30 to 13:39:05	0.16	0.08	0.07
8	3	No. 2, Heading	17,177	Exhaust Curtain	2	13:41:14 to 13:41:48	0.13	0.02	0.11
8	3	No. 2, Heading	17,177	Exhaust Curtain	1	13:43:30 to 13:44:15	0.12	0.03	0.09
8	3	No. 2, Heading	17,177	Exhaust Curtain	2	13:45:52 to 13:46:33	0.15	0.01	0.14
8	3	No. 2, Heading	17,177	Exhaust Curtain	1	13:48:15 to 13:48:50	0.10	0.04	0.06
8	3	No. 2, Heading	17,177	Exhaust Curtain	2	13:50:26 to 13:51:38	0.08	0.02	0.06
8	3	No. 2, Heading	17,177	Exhaust Curtain	1	13:53:51 to 13:54:41	0.07	0.10	0.00
8	3	No. 2, Heading	17,177	Exhaust Curtain	2	13:56:11 to 13:56:46	0.15	0.15	0.00
8	3	No. 2, Heading	17,177	Exhaust Curtain	1	13:58:38 to 13:59:14	0.15	0.04	0.10
8	3	No. 2, Heading	17,177	Exhaust Curtain	2	14:01:06 to 14:03:17	0.14	0.04	0.10
8	3	No. 2, Heading	17,177	Exhaust Curtain	1	14:04:59 to 14:05:33	0.10	0.02	0.08
8	3	No. 2, Heading	17,177	Exhaust Curtain	2	14:07:23 to 14:07:53	0.14	0.01	0.14
8	3	No. 2, Heading	17,177	Exhaust Curtain	1	14:09:35 to 14:10:11	0.12	0.01	0.12
8	3	No. 2, Heading	17,177	Exhaust Curtain	2	14:13:36 to 14:14:25	0.11	0.02	0.09
8	3	No. 2, Heading	17,177	Exhaust Curtain	1	14:16:01 to 14:16:37	0.14	0.01	0.13
8	3	No. 2, Heading	17,177	Exhaust Curtain	2	14:18:25 to 14:19:00	0.17	0.04	0.13
8	3	No. 2, Heading	17,177	Exhaust Curtain	1	14:20:56 to 14:22:35	0.19	0.04	0.15
8	3	No. 2, Heading	17,177	Exhaust Curtain	2	14:24:09 to 14:24:40	0.15	0.05	0.10
8	3	No. 2, Heading	17,177	Exhaust Curtain	1	14:26:17 to 14:27:08	0.13	0.03	0.09
8	3	No. 2, Heading	17,177	Exhaust Curtain	2	14:28:48 to 14:29:24	0.13	0.03	0.10

Table C- 4. Miner-generated dust levels during the regular- and deep-cut depths
Note: Dust levels are adjusted for productivity.

Cut No.	Cut Sequence	Depth	Curtain or Tubing Airflow (cfm)	Face Ventilation	Starting Scrubber Airflow (cfm)	Ending Scrubber Airflow (cfm)	Miner Intake Dust Level (mg/m^3)	Miner Return Dust Level (mg/m^3)	Miner Generated (mg/m^3)	Cars Per Minute	Adjusted Miner Generated (mg/m^3)
1	No. 1, Right	Regular	16,635	Exhausting	None	N/A	0.09	Invalid	Invalid	0.29	Invalid
1	No. 1, Right	Deep	16,635	Exhausting	N/A	12,700	0.10	Invalid	Invalid	0.28	Invalid
2	No. 2, Right	Regular	17,077	Exhausting	13,200	N/A	0.22	Invalid	Invalid	0.35	Invalid
2	No. 2, Right	Deep	17,077	Exhausting	N/A	12,900	0.07	Invalid	Invalid	0.38	Invalid
3	No. 4, Heading	Regular	21,522	Exhausting	13,300	N/A	0.06	1.62	1.56	0.47	1.32
3	No. 4, Heading	Deep	21,522	Exhausting	N/A	12,700	0.05	2.80	2.75	0.46	2.39
4	No. 5, Heading	Regular	16,743	Exhausting	13,700	N/A	0.07	1.68	1.62	0.49	1.33
4	No. 5, Heading	Deep	16,743	Exhausting	N/A	12,400	0.07	1.35	1.28	0.51	1.01
5	No. 4, Heading	Regular	18,325	Exhausting	14,000	N/A	0.08	1.30	1.21	0.32	1.54
5	No. 4, Heading	Deep	18,325	Exhausting	N/A	13,200	0.06	1.72	1.66	0.47	1.43
6	No. 4, Right	Regular	14,336	Exhausting	13,500	N/A	0.10	1.42	1.31	0.40	1.32
6	No. 4, Right	Deep	14,336	Exhausting	N/A	12,900	0.08	3.61	3.53	0.31	4.62
7	No. 3, Heading	Regular	17,261	Exhausting	14,000	N/A	0.13	1.69	1.56	0.19	3.34
7	No. 3, Heading	Deep	17,261	Exhausting	N/A	12,700	0.08	1.91	1.82	0.37	1.99
8	No. 2, Heading	Regular	17,177	Exhausting	14,200	N/A	0.10	2.02	1.92	0.43	1.79
8	No. 2, Heading	Deep	17,177	Exhausting	N/A	13,200	0.03	1.70	1.67	0.39	1.71

Table C- 5. Miner operator dust exposure levels net of intake dust level during the regular- and deep-cut depths
Note: Dust levels are adjusted for productivity.

Cut No.	Cut Sequence	Depth	Curtain or Tubing Airflow (cfm)	Face Ventilation	Starting Scrubber Airflow (cfm)	Ending Scrubber Airflow (cfm)	Miner Operator Exposure (mg/m^3)	Cars Per Minute	Adj. Miner Operator Exposure (mg/m^3)
1	No. 1, Right	Regular	16,635	Exhausting	None	N/A	0.58	0.29	0.79
1	No. 1, Right	Deep	16,635	Exhausting	N/A	12,700	0.91	0.28	1.32
2	No. 2, Right	Regular	17,077	Exhausting	13,200	N/A	0.74	0.35	0.84
2	No. 2, Right	Deep	17,077	Exhausting	N/A	12,900	0.44	0.38	0.47
3	No. 4, Heading	Regular	21,522	Exhausting	13,300	N/A	0.35	0.47	0.30
3	No. 4, Heading	Deep	21,522	Exhausting	N/A	12,700	0.66	0.46	0.57
4	No. 5, Heading	Regular	16,743	Exhausting	13,700	N/A	0.53	0.49	0.44
4	No. 5, Heading	Deep	16,743	Exhausting	N/A	12,400	0.62	0.51	0.49
5	No. 4, Heading	Regular	18,325	Exhausting	14,000	N/A	0.14	0.32	0.17
5	No. 4, Heading	Deep	18,325	Exhausting	N/A	13,200	0.43	0.47	0.37
6	No. 4, Right	Regular	14,336	Exhausting	13,500	N/A	0.08	0.40	0.08
6	No. 4, Right	Deep	14,336	Exhausting	N/A	12,900	0.74	0.31	0.97
7	No. 3, Heading	Regular	17,261	Exhausting	14,000	N/A	0.16	0.19	0.33
7	No. 3, Heading	Deep	17,261	Exhausting	N/A	12,700	0.82	0.37	0.89
8	No. 2, Heading	Regular	17,177	Exhausting	14,200	N/A	0.42	0.43	0.39
8	No. 2, Heading	Deep	17,177	Exhausting	N/A	13,200	0.44	0.39	0.45

Table C-6. Bolter machine-generated dust levels and bolter operator exposures for the regular- and deep-cut depths

Room No.	Entry	Depth	Position With Respect To Miner	Curtain Airflow (cfm)	Face Ventilation	Bolter Suction Left/Right (in Hg)	Left-Side Bolter Dust Levels mg/m3	Right-Side Bolter Dust Levels mg/m3	Upwind Bolter Dust Levels mg/m3	Downwind Bolter Dust Levels mg/m3	Bolter Generated Dust mg/m3
1	No. 1, Heading	Regular	Upwind	2600	Exhaust	N/A	0.55	Battery	0.42	Invalid	Invalid
1	No. 1, Heading	Deep	Upwind	2600	Exhaust	N/A	0.57	Battery	0.19	Invalid	Invalid
2	No. 2, Right	Regular	Upwind	3800	Exhaust	N/A	0.26	Battery	0.08	Invalid	Invalid
2	No. 2, Right	Deep	Upwind	3800	Exhaust	N/A	0.34	Battery	0.20	Invalid	Invalid
3	No. 5, Heading	Regular	Upwind	5000	Exhaust	N/A	0.15	0.17	0.04	Invalid	Invalid
3	No. 5, Heading	Deep	Upwind	5000	Exhaust	N/A	0.18	0.15	0.06	Invalid	Invalid
4	No. 4, Heading	Regular	Upwind	4600	Exhaust	N/A	0.16	0.16	0.26	Invalid	Invalid
4	No. 4, Heading	Deep	Upwind	4600	Exhaust	N/A	0.15	0.16	0.16	Invalid	Invalid
5	No. 5, Heading	Regular	Upwind	3300	Exhaust	N/A	0.14	0.14	0.17	Invalid	Invalid
5	No. 5, Heading	Deep	Upwind	3300	Exhaust	N/A	0.13	0.30	0.13	Invalid	Invalid
6	No. 4, Right	Regular	Upwind	14200	Exhaust	N/A	0.13	0.13	0.12	0.21	0.09
6	No. 4, Right	Deep	Upwind	14200	Exhaust	N/A	0.15	0.13	0.21	0.20	0.00
7	No. 3, Heading	Regular	Upwind	6400	Exhaust	N/A	0.26	0.13	0.08	0.13	0.05
7	No. 3, Heading	Deep	Upwind	6400	Exhaust	N/A	0.13	Battery	0.10	0.11	0.02

Appendix D: MINE D CASE STUDY

Mine-specific Information

Mine D used two Eimco 2810 continuous miners. The miner spray configuration is shown in Figure D-1. A total of 20 dust suppression sprays were operated at a pressure of not less than 75 psi as required by the mine ventilation plan—14 sprays at the top of the cutter drum, 4 pan sprays underneath the cutter drum (2 on each side of the miner), and 2 throat sprays. The ventilation plan required the scrubber capacity to be a minimum of 6,000 cfm using a 20-mesh scrubber filter. The plan also required the line curtain to be maintained within 50 ft of the deepest point of penetration with minimum required curtain airflows of 6000 cfm with the scrubber on and 5400 cfm with the scrubber off. Roof bolter faces required minimum curtain airflow of 3000 cfm.

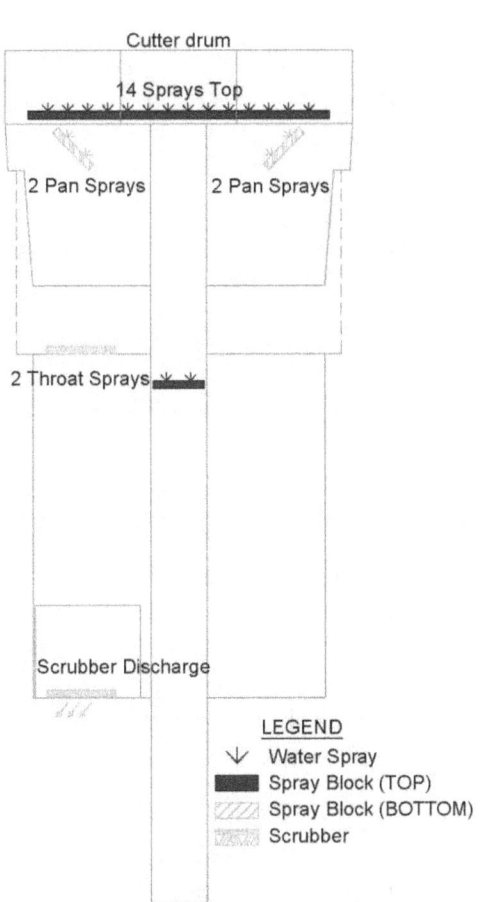

Figure D-1. Mine D continuous miner spray configuration.

The cut sequence for Mine D was as follows:

(1) 15-ft sump cut, right side
(2) 15-ft slab cut, left side

(3) 15-ft sump cut, left side
(4) 15-ft slab cut, right side

Figure D-2 shows a plan view of the continuous mining section on Day 1 of the study. The section was expanded to 10 entries on Day 2 and 11 entries on Day 3. The section used a sweeping ventilation configuration with Entries No. 8 and 9 serving as the intakes and Entries No. 1, 2, and 3 serving as the returns. All other entries were neutral, including the beltway, which was located in Entry No. 5. Mining height averaged 80-in, and entry width was 20 ft. Pillar dimensions were 30 x 40 ft. The depth of deep cuts were approximately 30 ft. Mine D is required to use line curtain when the distance from the point-of-origin to the coal face exceeds 50 ft. *Note: The point-of-origin is defined as the point at which the curtain will originate in order to direct the air to the face, usually the inby corner of the first outby pillar.* Due to the small pillar size used at Mine D, this distance was only exceeded once during the study. The Mine D dust survey was conducted on August 11, 12, and 13, 2009.

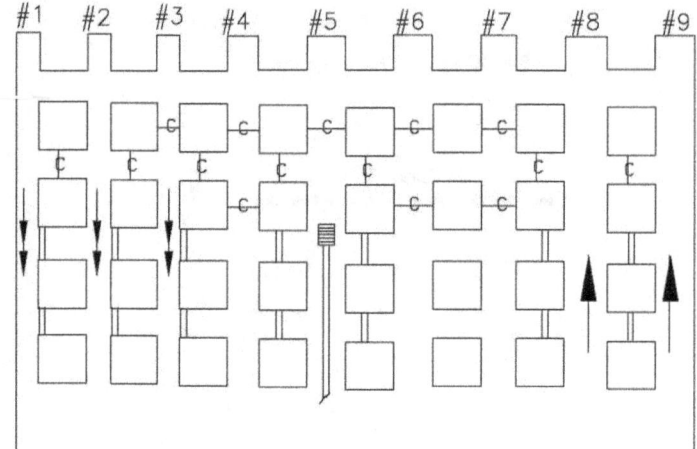

Figure D-2. Plan view of mine D continuous miner section.

Shuttle Car Data Analysis

Tables D-1, D-2, and D-3 show the shuttle car data collected during the study. Along with the two continuous miners, Mine D used three Narco shuttle cars on the section. The mine's ventilation plan requires that the continuous miners not be operated simultaneously. The survey was conducted on the right side of the section, where the miner was responsible for developing all entries to the right of the beltway. Two shuttle cars were operated behind this miner, and both had a standard cab configuration. Three unique types of deep-cuts were evaluated for this study:

(1) Continuous miner driving a heading without ventilation curtain
(2) Continuous miner turning a right crosscut without curtain
(3) Continuous miner driving a heading with blowing curtain

Figure D-3 is a plot of shuttle car dust exposures for six cuts (No. 2, 3, 4, 6, 7, and 8) where the miner was driving a heading without ventilation curtain. The x-axis corresponds to the sequential load number during the cut. A downward trend in shuttle car dust levels is evident during the development of each entry. This trend most likely resulted from shuttle car positioning with respect to ventilation airflow. For all of the no-curtain heading cuts, the left and right crosscuts were open to ventilation airflow, which was from right to left. During the initial sump, the miner's scrubber exhaust was outby the crosscuts, resulting in rollback into the shuttle car travel way. Ventilation airflow in the shuttle car travel way was not measureable for Cuts No. 2 and 6; away from the face for Cuts No. 3, 4, and 8; and toward the face for Cut No. 7. As the cut advanced, the scrubber exhaust was directed into the left crosscut, where it was quickly swept away from the workings. As the miner further advanced the heading, the shuttle car cabs traveled past the outby pillars, placing them directly into intake air. When driving a heading without the use of blowing curtain, shuttle car cab dust levels were 85% lower during the deep-cut sequence, averaging 0.62 mg/m^3 during the regular-cut depth and 0.09 mg/m^3 during the deep-cut depth. This result is significant when comparing the 95% CIs for the averages.

Figure D-4 is a plot of shuttle car dust exposures for two cuts (No. 1 and 5) where the miner was driving a right crosscut without curtain. These cuts are representative of blowing face ventilation because the scrubber exhaust coursed through the shuttle car travel way during the development of these entries. Again, a downward trend in shuttle car dust levels is evident. Lower shuttle car cab dust levels when mining the deep-cut portion of the right crosscuts are most likely explained by two factors. First, similar to the heading cuts without curtain, the shuttle car cab position improved with respect to the intake airflow as the crosscuts were advanced. In addition, miner dust generation levels averaged 66% lower (see "Miner-generated Dust") during the deep-cut depth when compared to the regular-cut depth during this survey, resulting in less dust in the shuttle car travel ways when mining the deep-cut portion of the right crosscuts. When developing a right crosscut without the use of blowing curtain, shuttle car cab dust levels were 59% lower (95% CIs) during the deep-cut sequence, averaging 0.74 mg/m^3 during the regular-cut depth and 0.30 mg/m^3 during deep-cut depth.

Figure D-5 is a plot of shuttle car dust exposures for Cut No. 9, where the miner was driving a heading with the use of blowing curtain. An upward trend in shuttle car cab dust levels is evident as the entry is advanced. An examination of face and scrubber airflow quantities may explain this increase. Airflow delivered to the face by the blowing curtain during this cut was approximately 9,300 cfm. The scrubber airflow was 7,300 cfm at the start of the cut, degrading to 5,100 cfm at the completion of the cut. Previous laboratory research demonstrated that a curtain-to-scrubber airflow ratio, before scrubber activation, of 1.0 is optimal for dust control on blowing curtain faces [Jayaraman et al. 1992]. The research also demonstrated that dust levels can increase significantly when either too much or too little air is delivered to the face. In this case, the airflow ratio, after scrubber activation, increased from 1.3 at the start of the cut to 1.8 by the end of the cut. Assuming the scrubber boosted airflow by 40%, a situation observed in a laboratory study at similar airflows [Taylor et al. 1997], the airflow ratio at Mine D, before scrubber activation, is estimated to have worsened from 0.9 to 1.3, which could have adversely impacted face area dust levels. Low ending scrubber airflow volumes may also have increased dust levels [Jayaraman et al. 1992]. When the miner was driving a heading with the use of blowing curtain, shuttle car cab dust levels were 72% higher (95% CIs) during the deep-cut sequence, averaging 0.75 mg/m^3 during the regular-cut depth and 2.73 mg/m^3 during the deep-cut depth.

The daily average dust levels for the shuttle car cabs were very low throughout the study, ranging from 0.11 to 0.53 mg/m^3, indicating that use of deep-cutting techniques did not hinder the mine's ability to limit the exposure levels of the shuttle car operators.

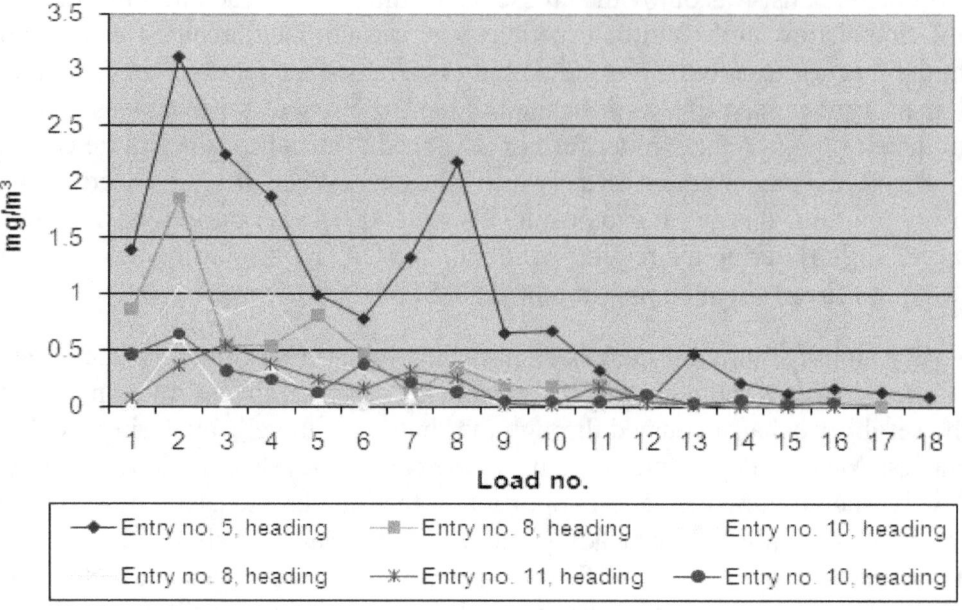

Figure D-3. Shuttle car dust levels at the face when the miner was driving headings without the use of line curtain.

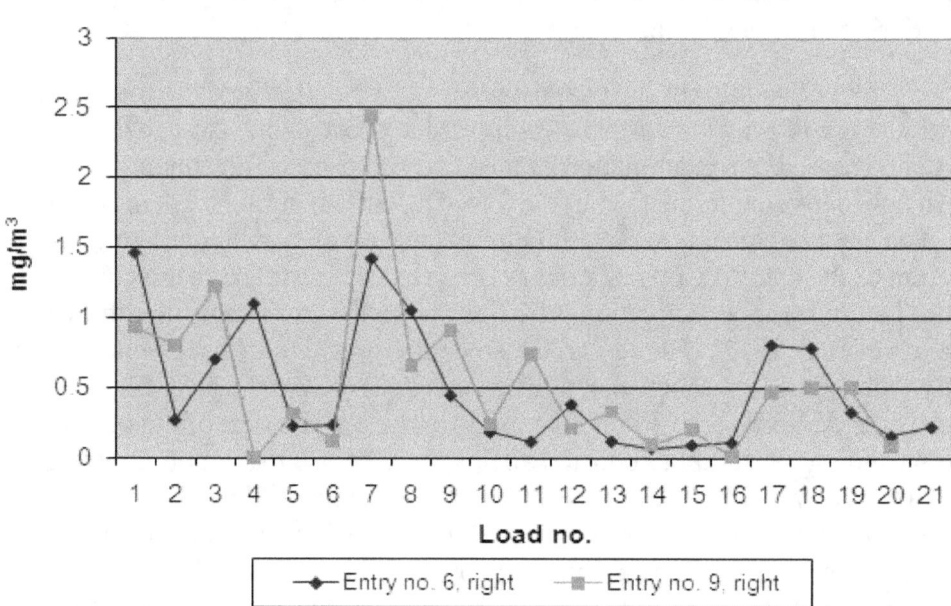

Figure D-4. Shuttle car dust levels at the face when the miner was driving right crosscuts without the use of curtain.

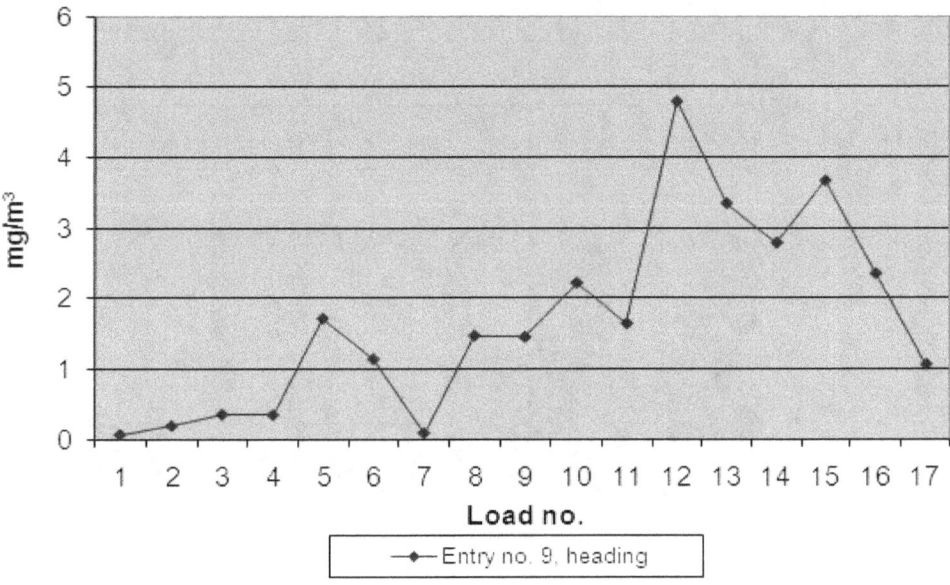

Figure D-5. Shuttle car dust levels at the face when the miner was driving a heading with the use of blowing curtain.

Miner-generated Dust

Table D-4 shows miner-generated dust levels during the regular- and deep-cut sequences for each room development. The last column of Table D-4 adjusts dust levels based on the measured productivity for each cut sequence, with the average number of cars loaded per minute being 0.44.

Miner-generated dust was lower during the deep-cut sequence when comparing it to the regular-cut sequence for each of the nine cuts surveyed during the study. In fact, miner-generated dust levels dropped significantly (95% CIs) from an average level of 2.69 mg/m^3 during the regular-cut depth to 0.91 mg/m^3 during the deep-cut depth (a 66% drop). Figure D-6 shows miner return dust levels at 5-sec intervals for each cut. Return dust levels are clearly higher during the initial sump and slab cuts, which are visually discernable on the graph. During the second sump and slab cuts, the dust levels are much lower and cannot be differentiated on the graph. This decrease may be due to the fact that airflow reaching the face is essentially limited to what the scrubber is drawing after the miner extends beyond the influence of the ventilation curtain or primary intake airflow. The resulting lower air velocities at the face lead to improved scrubber capture efficiencies.

The mine also provided an ample amount of air for dilution of miner-generated dust during the study, averaging 30,700 cfm. In general, miner-generated dust was very low for both the regular- and deep-cut sequences.

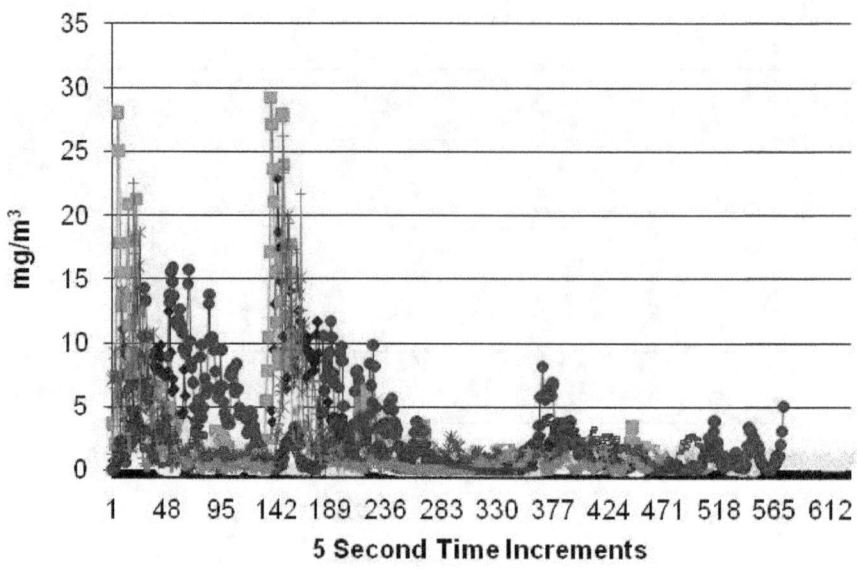

Figure D-6. Miner return dust levels at 5-sec intervals for each cut.

Miner Operator Data Analysis

The miner operator properly maintained his position in intake air throughout the study, resulting in extremely low dust exposure levels. Because levels were so low, no differentiation could be made between exposures during the deep- and regular-cut sequences. Day 3 miner operator exposure data were lost due to an improperly loaded filter cartridge in the PDM device. The daily average dust exposure levels for the miner operator were 0.08 mg/m^3 on Day 1 and 0.02 mg/m^3 on Day 2. Use of deep-cutting techniques did not hinder the mine's ability to limit the exposure levels of the miner operators.

Bolter Operations Data Analysis

Table D-5 summarizes the bolting sequences conducted during the study at Mine D. The mine used a dual-head Fletcher Roof Ranger II machine to perform roof bolting operations. The bolter operators maintained adequate blowing ventilation at the faces, averaging 5,800 cfm during the study. They also performed routine maintenance to the bolting machine's dust collection system, which assured proper suction pressure for dust capture. All bolting operations were conducted upwind of mining operations. As a result of these conditions, the bolter operators' daily average dust exposure levels were extremely low during the study, ranging between 0.05 and 0.22 mg/m^3. Because levels were so low, no differentiation could be made between exposures during the deep- and regular-cut sequences. Use of deep-cutting techniques did not hinder the mine's ability to limit the exposure levels of the bolter operators.

Dust-control Monitoring

As mentioned earlier, one of the concerns was that the scrubber would become clogged with particles as the mining cycle progressed through the deep cut, diminishing its effectiveness. Table D-4 shows scrubber airflows at the start and end of each cut. A 26% average drop-off in scrubber airflow was observed from the start to the end of the cuts. However, this decrease in scrubber airflow did not produce a negative effect on downwind dust levels at Mine D, nor did it appear to affect the dust exposure levels of the face workers. Two factors may have contributed to this observation. First, as discussed in the "Miner-generated Dust" section of this report, dust capture appears to improve as the cutting depth increases beyond the influence of face ventilation airflow. In addition, higher pressure drops across the loaded scrubber filters may be improving scrubber efficiencies associated with respirable-sized particles enough to somewhat compensate for the processing of lower volumes of air.

As with the continuous miner, no degradation in the bolting machine's dust control circuit was observed during completion of the bolting sequence in the deep-cut rooms. Suction measurements were taken at the start, middle, and end of each bolting sequence, and all measurements were between 16 and 18-in Hg, which provided excellent dust control.

Summary

1. When driving a heading without the use of blowing curtain, shuttle car cab dust levels were 85% lower during the deep-cut sequence, averaging 0.62 mg/m^3 during the regular-cut depth and 0.09 mg/m^3 during the deep-cut depth.

2. When developing a right crosscut without the use of blowing curtain, shuttle car cab dust levels were 59% lower during the deep-cut sequence, averaging 0.74 mg/m^3 during the regular-cut depth and 0.30 mg/m^3 during the deep-cut depth.

3. When the miner was driving a heading with the use of blowing curtain, shuttle car cab dust levels were 72% higher during the deep-cut sequence, averaging 0.75 mg/m^3 during the regular-cut depth and 2.73 mg/m^3 during the deep-cut depth. This may have resulted from a less than ideal ending curtain-to-scrubber airflow ratio or due to low ending scrubber airflow.

4. The daily average dust levels for the shuttle car cabs were very low throughout the study, ranging from 0.11 to 0.53 mg/m^3, indicating that use of deep-cutting techniques did not hinder the mine's ability to limit the exposure levels of the shuttle car operators.

5. Miner-generated dust levels averaged 66% lower during the deep-cut sequence when comparing it to the regular-cut sequence.

6. Use of deep-cutting techniques did not hinder the mine's ability to limit the exposure levels of the miner operator, which were very low (< 0.10 mg/m^3) throughout the study.

7. Use of deep-cutting techniques did not hinder the mine's ability to limit the exposure levels of the bolter operators, which were very low (0.05 to 0.22 mg/m^3) throughout the study.

8. A 26% average drop-off in scrubber airflow was observed from the start to the end of the cuts. However, this decrease in scrubber airflow did not produce a negative effect on downwind dust levels at Mine D nor did it appear to affect the dust exposure levels of the face workers.

References Cited in Appendix D

Jayaraman NI, Jankowski RA, Whitehead KL [1992]. Optimizing continuous miner scrubbers for dust control in high coal seams. In: Proceedings of New Technology in Mine Health and Safety, SME Annual Meeting. Littleton, CO: Society for Mining, Metallurgy, and Exploration, pp. 193–205.

Taylor DT, Rider JP, Thimons ED [1997]. Impact of unbalanced intake and scrubber flows on face methane concentrations. In: Ramani RV, ed. Proceedings of the 6th International Mine Ventilation Congress. Chapter 27. Littleton, CO: Society for Mining, Metallurgy, and Exploration, pp. 169–172.

Table D-1. Shuttle car cab dust levels when loading at the face during Day 1 of the study
Note: Shaded areas show deep-cut depth, and clear areas show regular-cut depth.

Cut No.	Day	Cut Sequence	Curtain Airflow	Face Ventilation	Car No.	Time at Face	Shuttle Car Dust (mg/m^3)	Intake Dust Level (mg/m^3)	Adjusted Shuttle Car Dust (mg/m^3)
1	1	No. 6, Right	0	No Curtain	2	8:28:51 to 8:30:09	1.50	0.05	1.46
1	1	No. 6, Right	0	No Curtain	2	8:32:30 to 8:33:23	0.33	0.06	0.27
1	1	No. 6, Right	0	No Curtain	1	8:34:36 to 8:35:36	0.77	0.07	0.70
1	1	No. 6, Right	0	No Curtain	2	8:36:09 to 8:36:58	1.16	0.06	1.10
1	1	No. 6, Right	0	No Curtain	1	8:38:04 to 8:38:50	0.29	0.07	0.23
1	1	No. 6, Right	0	No Curtain	2	8:39:28 to 8:40:33	0.30	0.06	0.23
1	1	No. 6, Right	0	No Curtain	1	8:42:00 to 8:43:30	1.47	0.06	1.42
1	1	No. 6, Right	0	No Curtain	2	8:44:00 to 8:45:15	1.10	0.05	1.05
1	1	No. 6, Right	0	No Curtain	1	8:45:54 to 8:46:55	0.51	0.06	0.44
1	1	No. 6, Right	0	No Curtain	2	8:47:27 to 8:48:40	0.23	0.05	0.18
1	1	No. 6, Right	0	No Curtain	1	8:55:30 to 8:56:33	0.17	0.05	0.12
1	1	No. 6, Right	0	No Curtain	2	8:57:12 to 8:58:15	0.43	0.05	0.38
1	1	No. 6, Right	0	No Curtain	1	8:59:00 to 8:59:56	0.17	0.06	0.12
1	1	No. 6, Right	0	No Curtain	2	9:00:37 to 9:01:38	0.12	0.05	0.07
1	1	No. 6, Right	0	No Curtain	1	9:02:35 to 9:03:32	0.14	0.04	0.09
1	1	No. 6, Right	0	No Curtain	2	9:04:16 to 9:05:53	0.14	0.03	0.11
1	1	No. 6, Right	0	No Curtain	1	9:07:50 to 9:09:07	0.84	0.03	0.81
1	1	No. 6, Right	0	No Curtain	2	9:09:51 to 9:10:52	0.82	0.03	0.78
1	1	No. 6, Right	0	No Curtain	1	9:11:45 to 9:12:46	0.37	0.04	0.33
1	1	No. 6, Right	0	No Curtain	2	9:13:30 to 9:14:23	0.21	0.05	0.16
1	1	No. 6, Right	0	No Curtain	1	9:15:26 to 9:16:40	0.28	0.06	0.22
2	1	No. 5, Heading	0	No Curtain	2	10:33:42 to 10:35:23	1.43	0.04	1.39
2	1	No. 5, Heading	0	No Curtain	1	10:35:42 to 10:36:42	3.14	0.03	3.11
2	1	No. 5, Heading	0	No Curtain	2	10:37:07 to 10:38:13	2.29	0.04	2.24
2	1	No. 5, Heading	0	No Curtain	1	10:38:49 to 10:39:34	1.89	0.02	1.87
2	1	No. 5, Heading	0	No Curtain	2	10:39:56 to 10:41:00	1.02	0.03	0.99
2	1	No. 5, Heading	0	No Curtain	1	10:41:41 to 10:42:35	0.82	0.04	0.78
2	1	No. 5, Heading	0	No Curtain	2	10:44:52 to 10:46:27	1.36	0.03	1.33
2	1	No. 5, Heading	0	No Curtain	1	10:46:45 to 10:48:09	2.22	0.04	2.17
2	1	No. 5, Heading	0	No Curtain	2	10:48:36 to 10:49:33	0.69	0.04	0.65
2	1	No. 5, Heading	0	No Curtain	1	10:50:14 to 10:51:04	0.71	0.04	0.67
2	1	No. 5, Heading	0	No Curtain	2	10:51:40 to 10:52:40	0.36	0.03	0.32
2	1	No. 5, Heading	0	No Curtain	2	10:58:33 to 11:00:04	0.06	0.03	0.03
2	1	No. 5, Heading	0	No Curtain	1	11:00:36 to 11:01:40	0.49	0.03	0.47
2	1	No. 5, Heading	0	No Curtain	2	11:02:06 to 11:03:13	0.25	0.04	0.21
2	1	No. 5, Heading	0	No Curtain	1	11:03:50 to 11:05:04	0.16	0.04	0.12
2	1	No. 5, Heading	0	No Curtain	2	11:06:31 to 11:07:43	0.21	0.04	0.16
2	1	No. 5, Heading	0	No Curtain	1	11:08:08 to 11:09:12	0.16	0.03	0.13
2	1	No. 5, Heading	0	No Curtain	2	11:09:36 to 11:10:37	0.12	0.03	0.09
3	1	No. 8, Heading	0	No Curtain	2	12:25:11 to 12:26:52	0.92	0.07	0.86
3	1	No. 8, Heading	0	No Curtain	1	12:27:35 to 12:28:32	1.93	0.07	1.85
3	1	No. 8, Heading	0	No Curtain	2	12:29:25 to 12:30:24	0.62	0.08	0.53
3	1	No. 8, Heading	0	No Curtain	1	12:31:01 to 12:32:00	0.62	0.09	0.54
3	1	No. 8, Heading	0	No Curtain	2	12:32:40 to 12:34:40	0.90	0.09	0.81
3	1	No. 8, Heading	0	No Curtain	1	12:36:25 to 12:37:29	0.56	0.09	0.47
3	1	No. 8, Heading	0	No Curtain	2	12:42:26 to 12:43:08	0.21	0.10	0.11
3	1	No. 8, Heading	0	No Curtain	1	12:43:46 to 12:44:54	0.45	0.10	0.35
3	1	No. 8, Heading	0	No Curtain	2	12:45:31 to 12:46:39	0.29	0.10	0.18
3	1	No. 8, Heading	0	No Curtain	1	12:51:54 to 12:53:08	0.26	0.09	0.17
3	1	No. 8, Heading	0	No Curtain	2	12:53:52 to 12:55:02	0.29	0.09	0.20
3	1	No. 8, Heading	0	No Curtain	1	12:55:57 to 12:57:10	0.16	0.10	0.06
3	1	No. 8, Heading	0	No Curtain	2	12:57:52 to 12:58:53	0.12	0.10	0.02
3	1	No. 8, Heading	0	No Curtain	1	12:59:37 to 13:00:31	0.11	0.09	0.02
3	1	No. 8, Heading	0	No Curtain	2	13:02:25 to 13:04:09	0.10	0.09	0.01
3	1	No. 8, Heading	0	No Curtain	1	13:04:55 to 13:06:10	0.13	0.09	0.04
3	1	No. 8, Heading	0	No Curtain	2	13:06:51 to 13:07:50	0.09	0.09	0.00

Table D-2. Shuttle car cab dust levels when loading at the face during Day 2 of the study
Note: Shaded areas show deep-cut depth, and clear areas show regular-cut depth.

Cut No.	Day	Cut Sequence	Curtain Airflow	Face Ventilation	Car No.	Time at Face	Shuttle Car Dust (mg/m^3)	Intake Dust Level (mg/m^3)	Adjusted Shuttle Car Dust (mg/m^3)
4	2	No. 10, Heading	0	No Curtain	1	8:31:31 to 8:32:18	0.06	0.04	0.02
4	2	No. 10, Heading	0	No Curtain	2	8:33:18 to 8:34:17	0.66	0.04	0.62
4	2	No. 10, Heading	0	No Curtain	1	8:35:38 to 8:37:14	0.11	0.05	0.06
4	2	No. 10, Heading	0	No Curtain	2	8:38:03 to 8:39:02	0.40	0.04	0.35
4	2	No. 10, Heading	0	No Curtain	1	8:39:55 to 8:41:18	0.11	0.05	0.07
4	2	No. 10, Heading	0	No Curtain	2	8:42:43 to 8:43:33	0.08	0.05	0.02
4	2	No. 10, Heading	0	No Curtain	1	8:44:22 to 8:45:13	0.15	0.06	0.09
4	2	No. 10, Heading	0	No Curtain	2	8:46:02 to 8:47:02	0.21	0.05	0.16
4	2	No. 10, Heading	0	No Curtain	1	8:47:52 to 8:48:52	0.07	0.05	0.01
4	2	No. 10, Heading	0	No Curtain	2	8:58:00 to 8:59:40	0.08	0.05	0.03
4	2	No. 10, Heading	0	No Curtain	1	9:00:29 to 9:01:58	0.08	0.05	0.03
4	2	No. 10, Heading	0	No Curtain	2	9:03:01 to 9:04:17	0.10	0.05	0.05
4	2	No. 10, Heading	0	No Curtain	1	9:05:10 to 9:06:38	0.03	0.05	0.00
4	2	No. 10, Heading	0	No Curtain	2	9:08:29 to 9:09:45	0.13	0.05	0.08
4	2	No. 10, Heading	0	No Curtain	1	9:10:40 to 9:11:30	0.07	0.05	0.02
4	2	No. 10, Heading	0	No Curtain	2	9:12:24 to 9:13:35	0.07	0.06	0.01
5	2	No. 9, Right	0	No Curtain	2	10:05:45 to 10:07:14	1.34	0.41	0.93
5	2	No. 9, Right	0	No Curtain	1	10:07:35 to 10:09:01	0.96	0.16	0.80
5	2	No. 9, Right	0	No Curtain	2	10:09:24 to 10:11:08	1.36	0.14	1.22
5	2	No. 9, Right	0	No Curtain	1	10:11:36 to 10:13:14	0.36	0.36	0.00
5	2	No. 9, Right	0	No Curtain	2	10:13:44 to 10:15:10	0.47	0.15	0.32
5	2	No. 9, Right	0	No Curtain	1	10:15:32 to 10:17:40	0.27	0.15	0.12
5	2	No. 9, Right	0	No Curtain	2	10:23:43 to 10:25:17	2.52	0.08	2.44
5	2	No. 9, Right	0	No Curtain	1	10:26:00 to 10:27:04	0.78	0.12	0.66
5	2	No. 9, Right	0	No Curtain	2	10:27:39 to 10:28:39	0.99	0.08	0.91
5	2	No. 9, Right	0	No Curtain	1	10:37:07 to 10:37:40	0.26	0.02	0.24
5	2	No. 9, Right	0	No Curtain	2	10:38:19 to 10:39:11	0.77	0.02	0.74
5	2	No. 9, Right	0	No Curtain	1	10:39:53 to 10:40:54	0.24	0.03	0.21
5	2	No. 9, Right	0	No Curtain	2	10:41:41 to 10:42:52	0.37	0.04	0.33
5	2	No. 9, Right	0	No Curtain	1	10:43:32 to 10:44:36	0.12	0.03	0.10
5	2	No. 9, Right	0	No Curtain	2	10:45:23 to 10:46:54	0.23	0.03	0.20
5	2	No. 9, Right	0	No Curtain	1	10:47:39 to 10:49:15	0.08	0.06	0.02
5	2	No. 9, Right	0	No Curtain	2	10:51:35 to 10:52:50	0.51	0.04	0.47
5	2	No. 9, Right	0	No Curtain	1	10:53:38 to 10:54:20	0.54	0.03	0.50
5	2	No. 9, Right	0	No Curtain	2	10:55:08 to 10:55:59	0.53	0.03	0.50
5	2	No. 9, Right	0	No Curtain	1	10:56:50 to 10:58:04	0.12	0.04	0.08
6	2	No. 8, Heading	0	No Curtain	1	11:39:52 to 11:40:55	0.78	0.37	0.41
6	2	No. 8, Heading	0	No Curtain	2	11:41:15 to 11:42:20	1.22	0.16	1.06
6	2	No. 8, Heading	0	No Curtain	1	11:42:50 to 11:43:52	0.90	0.09	0.82
6	2	No. 8, Heading	0	No Curtain	2	11:44:13 to 11:45:15	1.07	0.09	0.98
6	2	No. 8, Heading	0	No Curtain	1	11:45:50 to 11:46:52	0.50	0.11	0.39
6	2	No. 8, Heading	0	No Curtain	2	11:47:19 to 11:48:59	0.47	0.16	0.30
6	2	No. 8, Heading	0	No Curtain	1	11:51:23 to 11:52:38	0.26	0.23	0.03
6	2	No. 8, Heading	0	No Curtain	2	11:53:10 to 11:54:03	0.57	0.19	0.38
6	2	No. 8, Heading	0	No Curtain	1	11:54:35 to 11:55:39	0.36	0.11	0.24
6	2	No. 8, Heading	0	No Curtain	2	11:56:11 to 11:57:19	0.23	0.13	0.10
6	2	No. 8, Heading	0	No Curtain	1	12:02:55 to 12:04:16	0.08	0.08	0.00
6	2	No. 8, Heading	0	No Curtain	2	12:04:40 to 12:05:46	0.25	0.10	0.15
6	2	No. 8, Heading	0	No Curtain	1	12:06:22 to 12:07:26	0.06	0.09	0.00
6	2	No. 8, Heading	0	No Curtain	2	12:08:51 to 12:10:17	0.09	0.05	0.04
6	2	No. 8, Heading	0	No Curtain	1	12:10:49 to 12:11:51	0.09	0.16	0.00

Table D-3. Shuttle car cab dust levels when loading at the face during Day 3 of the study
Note: Shaded areas show deep-cut depth, and clear areas show regular-cut depth.
Curtain airflow was measured after activation of the scrubber.

Cut No.	Day	Cut Sequence	Curtain Airflow	Face Ventilation	Car No.	Time at Face	Shuttle Car Dust (mg/m³)	Intake Dust Level (mg/m³)	Adjusted Shuttle Car Dust (mg/m³)
7	3	No. 11, Heading	0	No Curtain	2	8:52:14 to 8:53:15	0.11	0.05	0.07
7	3	No. 11, Heading	0	No Curtain	1	8:55:35 to 8:56:54	0.41	0.05	0.36
7	3	No. 11, Heading	0	No Curtain	2	8:58:04 to 8:59:05	0.61	0.05	0.55
7	3	No. 11, Heading	0	No Curtain	1	9:00:12 to 9:01:19	0.44	0.06	0.38
7	3	No. 11, Heading	0	No Curtain	1	9:04:06 to 9:05:16	0.31	0.07	0.24
7	3	No. 11, Heading	0	No Curtain	2	9:07:09 to 9:08:11	0.23	0.07	0.16
7	3	No. 11, Heading	0	No Curtain	1	9:09:15 to 9:10:13	0.39	0.08	0.31
7	3	No. 11, Heading	0	No Curtain	2	9:11:13 to 9:12:07	0.33	0.08	0.25
7	3	No. 11, Heading	0	No Curtain	1	9:13:11 to 9:14:11	0.10	0.08	0.02
7	3	No. 11, Heading	0	No Curtain	2	9:21:23 to 9:22:42	0.11	0.10	0.01
7	3	No. 11, Heading	0	No Curtain	1	9:23:52 to 9:25:07	0.27	0.10	0.17
7	3	No. 11, Heading	0	No Curtain	2	9:26:13 to 9:27:12	0.14	0.11	0.03
7	3	No. 11, Heading	0	No Curtain	1	9:28:25 to 9:29:25	0.12	0.10	0.01
7	3	No. 11, Heading	0	No Curtain	2	9:33:26 to 9:34:56	0.10	0.11	0.00
7	3	No. 11, Heading	0	No Curtain	1	9:36:45 to 9:37:47	0.05	0.10	0.00
7	3	No. 11, Heading	0	No Curtain	2	9:38:50 to 9:39:53	0.07	0.10	0.00
8	3	No. 10, Heading	0	No Curtain	1	10:37:47 to 10:38:52	0.52	0.07	0.46
8	3	No. 10, Heading	0	No Curtain	2	10:40:02 to 10:40:51	0.68	0.04	0.64
8	3	No. 10, Heading	0	No Curtain	1	10:41:52 to 10:43:01	0.36	0.04	0.32
8	3	No. 10, Heading	0	No Curtain	2	10:43:53 to 10:45:15	0.27	0.03	0.24
8	3	No. 10, Heading	0	No Curtain	1	10:46:13 to 10:47:26	0.15	0.03	0.12
8	3	No. 10, Heading	0	No Curtain	2	10:49:43 to 10:50:51	0.41	0.03	0.38
8	3	No. 10, Heading	0	No Curtain	1	10:51:44 to 10:52:48	0.24	0.03	0.21
8	3	No. 10, Heading	0	No Curtain	2	10:53:45 to 10:54:39	0.16	0.03	0.13
8	3	No. 10, Heading	0	No Curtain	1	10:55:37 to 10:56:30	0.08	0.03	0.05
8	3	No. 10, Heading	0	No Curtain	2	11:03:02 to 11:04:02	0.08	0.03	0.05
8	3	No. 10, Heading	0	No Curtain	1	11:05:06 to 11:06:01	0.08	0.03	0.05
8	3	No. 10, Heading	0	No Curtain	2	11:06:56 to 11:07:55	0.14	0.03	0.11
8	3	No. 10, Heading	0	No Curtain	1	11:08:58 to 11:10:03	0.06	0.03	0.03
8	3	No. 10, Heading	0	No Curtain	2	11:11:34 to 11:12:48	0.09	0.03	0.06
8	3	No. 10, Heading	0	No Curtain	1	11:13:55 to 11:14:51	0.05	0.03	0.02
8	3	No. 10, Heading	0	No Curtain	2	11:15:52 to 11:16:55	0.07	0.03	0.04
9	3	No. 9, Heading	9,300	Blowing	1	12:09:03 to 12:10:19	0.18	0.12	0.06
9	3	No. 9, Heading	9,300	Blowing	2	12:12:16 to 12:13:14	0.21	0.03	0.19
9	3	No. 9, Heading	9,300	Blowing	1	12:14:00 to 12:15:02	0.38	0.03	0.35
9	3	No. 9, Heading	9,300	Blowing	2	12:15:48 to 12:16:33	0.37	0.03	0.34
9	3	No. 9, Heading	9,300	Blowing	1	12:17:43 to 12:19:04	1.74	0.03	1.71
9	3	No. 9, Heading	9,300	Blowing	2	12:19:56 to 12:21:26	1.16	0.03	1.13
9	3	No. 9, Heading	9,300	Blowing	1	12:24:08 to 12:25:10	0.11	0.03	0.08
9	3	No. 9, Heading	9,300	Blowing	2	12:25:55 to 12:26:55	1.49	0.03	1.46
9	3	No. 9, Heading	9,300	Blowing	1	12:27:46 to 12:29:03	1.47	0.03	1.45
9	3	No. 9, Heading	9,300	Blowing	2	12:29:50 to 12:30:59	2.25	0.03	2.22
9	3	No. 9, Heading	9,300	Blowing	1	12:38:04 to 12:39:24	1.66	0.03	1.64
9	3	No. 9, Heading	9,300	Blowing	2	12:40:17 to 12:41:08	4.82	0.03	4.79
9	3	No. 9, Heading	9,300	Blowing	1	12:41:57 to 12:43:35	3.37	0.03	3.34
9	3	No. 9, Heading	9,300	Blowing	2	12:44:27 to 12:45:24	2.81	0.03	2.78
9	3	No. 9, Heading	9,300	Blowing	1	12:48:31 to 12:49:43	3.69	0.03	3.66
9	3	No. 9, Heading	9,300	Blowing	2	12:50:37 to 12:51:39	2.38	0.03	2.35
9	3	No. 9, Heading	9,300	Blowing	1	12:52:34 to 12:53:00	1.09	0.03	1.06

Table D-4. Miner-generated dust levels during the regular- and deep-cut depths
Note: Dust levels are adjusted for productivity.

Cut No.	Cut Sequence	Depth	Entry Airflow (cfm)	Blowing Curtain Airflow (cfm)	Starting Scrubber Airflow (cfm)	Ending Scrubber Airflow (cfm)	Miner Intake Dust Level (mg/m^3)	Miner Return Dust Level (mg/m^3)	Miner Generated (mg/m^3)	Cars Per Minute	Miner Generated Adjusted (mg/m^3)
1	No. 6, Right	Regular	21,100	0	7,200	N/A	0.06	0.69	0.63	0.50	0.55
1	No. 6, Right	Deep	21,100	0	N/A	4,700	0.05	0.28	0.24	0.52	0.20
2	No. 5, Heading	Regular	22,100	0	7,000	N/A	0.03	5.18	5.15	0.57	3.97
2	No. 5, Heading	Deep	22,100	0	N/A	4,600	0.04	1.08	1.05	0.44	1.05
3	No. 8, Heading	Regular	50,100	0	7,100	N/A	0.09	4.44	4.35	0.42	4.56
3	No. 8, Heading	Deep	50,100	0	N/A	4,800	0.09	0.77	0.68	0.50	0.60
4	No. 10, Heading	Regular	40,800	0	7,200	N/A	0.05	1.87	1.82	0.52	1.54
4	No. 10, Heading	Deep	40,800	0	N/A	6,000	0.05	0.53	0.48	0.31	0.68
5	No. 9, Right	Regular	27,600	0	7,100	N/A	0.12	1.18	1.06	0.31	1.50
5	No. 9, Right	Deep	27,600	0	N/A	5,100	0.03	0.83	0.80	0.51	0.69
6	No. 8, Heading	Regular	36,900	0	6,900	N/A	0.16	3.47	3.32	0.55	2.66
6	No. 8, Heading	Deep	36,900	0	N/A	5,700	0.09	1.39	1.30	0.43	1.33
7	No. 11, Heading	Regular	31,900	0	7,100	N/A	0.06	4.68	4.62	0.40	5.08
7	No. 11, Heading	Deep	31,900	0	N/A	5,400	0.10	1.10	1.00	0.30	1.46
8	No. 10, Heading	Regular	36,200	0	7,800	N/A	0.03	2.83	2.79	0.47	2.62
8	No. 10, Heading	Deep	36,200	0	N/A	6,300	0.03	0.48	0.45	0.38	0.52
9	No. 9, Heading	Regular	23,700	9,300	7,300	N/A	0.03	1.81	1.77	0.45	1.73
9	No. 9, Heading	Deep	23,700	9,300	N/A	5,100	0.03	1.36	1.33	0.35	1.67

Table D-5. Bolter-generated dust levels and operator exposures for each bolting sequence

Room No.	Entry	Depth	Position With Respect To Miner	Curtain Airflow (cfm)	Face Ventilation	Bolter Suction Left/Right (in Hg)	Left-Side Bolter Dust Levels (mg/m^3)	Right-Side Bolter Dust Levels (mg/m^3)	Upwind Bolter Dust Levels (mg/m^3)	Downwind Bolter Dust Levels (mg/m^3)	Bolter Generated Dust (mg/m^3)
1	No. 6, Right	Regular	Upwind	5,200	Blowing	17/17	0.00	0.00	0.11	0.09	0.00
1	No. 6, Right	Deep	Upwind	5,200	Blowing	17/17	0.24	0.24	0.15	0.12	0.00
2	No. 5, Heading	Regular	Upwind	3,700	Blowing	17/17	0.17	0.16	0.02	0.11	0.09
2	No. 5, Heading	Deep	Upwind	3,700	Blowing	17/17	0.00	0.00	0.02	0.20	0.18
3	No. 8, Heading	Regular	Upwind	9,000	Blowing	18/18	0.00	0.00	0.24	0.12	0.00
3	No. 8, Heading	Deep	Upwind	9,000	Blowing	18/18	0.00	0.00	0.37	0.08	0.00
4	No. 10, Heading	Regular	Upwind	8,600	Blowing	17/17	0.00	0.00	0.02	0.09	0.07
4	No. 10, Heading	Deep	Upwind	8,600	Blowing	17/17	0.00	0.00	0.03	0.16	0.13
5	No. 9, Right	Regular	Upwind	2,600	Blowing	18/17	0.00	0.11	0.03	0.12	0.09
5	No. 9, Right	Deep	Upwind	2,600	Blowing	18/17	0.00	0.00	0.05	0.13	0.08
6	No. 8, Heading	Regular	Upwind	7,600	Blowing	17/17	0.18	0.00	0.08	0.11	0.03
6	No. 8, Heading	Deep	Upwind	7,600	Blowing	17/17	0.00	0.36	0.12	0.10	0.00
7	No. 6, Heading	Regular	Upwind	2,000	Blowing	16/17	0.00	0.15	0.02	0.15	0.13
7	No. 6, Heading	Deep	Upwind	2,000	Blowing	16/17	0.35	0.00	0.07	0.22	0.15
8	No. 11, Heading	Regular	Upwind	7,000	Blowing	17/18	0.00	0.00	0.02	0.08	0.06
8	No. 11, Heading	Deep	Upwind	7,000	Blowing	17/18	0.00	0.00	0.05	0.09	0.05
9	No. 10, Heading	Regular	Upwind	7,000	Blowing	16/17	0.22	0.00	0.16	0.16	0.00
9	No. 10, Heading	Deep	Upwind	7,000	Blowing	16/17	0.00	0.40	0.25	0.15	0.00

Appendix E: MINE E CASE STUDY

Mine-specific Information

Mine E used two Joy 12C continuous miners. The miner spray configuration is shown in Figure E-1. A total of 46 dust suppression sprays (48 if using spray fan) were operated at a pressure of not less than 100 psi—15 sprays at the top of the cutter drum, 10 pan sprays underneath the cutter drum (5 on each side of the miner), 3 throat sprays, 12 side sprays (6 on each side of the miner), and 6 gathering head sprays behind the cutter motor (3 on each side of the miner). The mine ventilation plan only required 39 sprays to be operational. The ventilation plan required the scrubber capacity to be a minimum of 5,000 cfm. The plan also required the line curtain to be maintained within 40 ft of the deepest point of penetration and the minimum face curtain airflow was 6000 cfm.

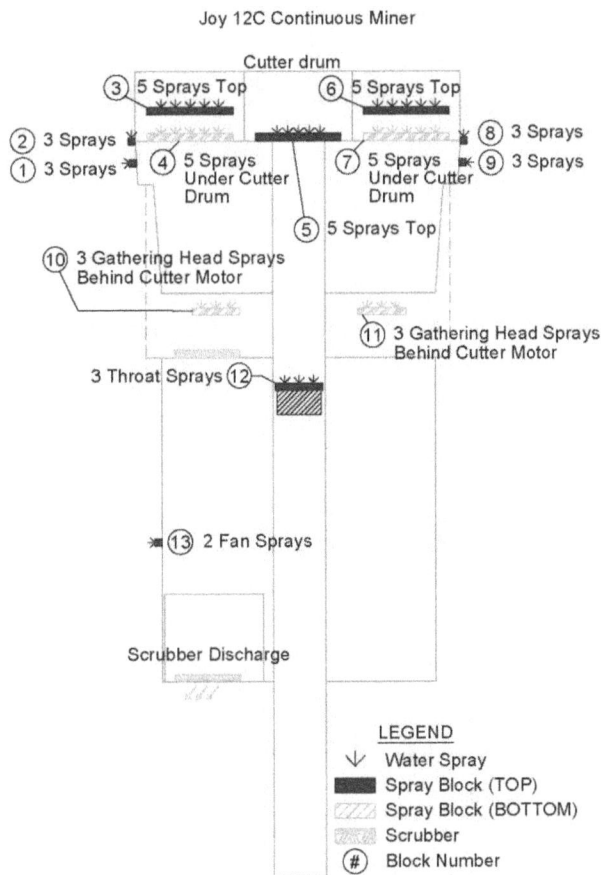

Figure E-1. Mine E continuous miner spray configuration.

Figure E-2 shows a plan view of the main entries for the study at Mine E. The mine used a split ventilation configuration to develop the mains, with Entries No. 6 and 7 serving as the intakes and Entries No. 1, 2, 8, and 9 serving as the returns. All other entries were neutral, including the beltway, which was located in Entry No. 5. Only Cuts No. 9 and 10 were conducted in the mains. The first eight cuts of the study were driven in three satellite entries driven off the right side of the section perpendicular to Entry No. 9, as shown in Figure E-3. These satellite entries were ventilated using a sweeping configuration, with Entry No. S1 serving as the intake, Entry No. S2 serving as the neutral, and Entry No. S3 serving as the return. The mining height throughout the study averaged 7 ft, with entry widths maintained at 20 ft. Pillar dimensions were 40 x 40 ft. The deep-cut depth ranged from 32 to 40 ft. Several cuts mined on Day 2 of the study were not analyzed because they were less than 30 ft in depth. The mine had a maximum curtain setback distance of 40 ft; however, due to small pillar dimensions, several cuts were mined without the use of curtain. The Mine E dust survey was conducted over 3 consecutive days from October 20 through 22, 2009.

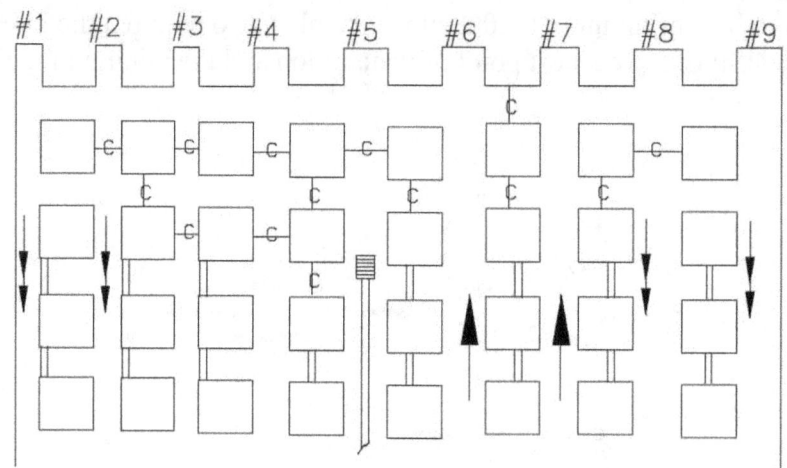

Figure E-2. Plan view of the mine E continuous miner Section No. 4.

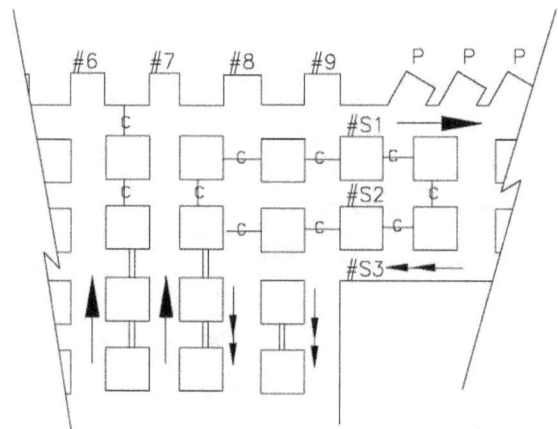

Figure E-3. Plan view of the satellite entries driven off the mains on Section No. 4.

Shuttle Car Data Analysis

Tables E-1, E-2, and E-3 show the shuttle car data collected during the study. Testing was completed on one continuous miner with three to four battery-powered, 10.5-ton Stamler shuttle cars for each cut. All shuttle cars had a standard cab configuration, placing the operators on the off-curtain side of the entry. Four unique types of deep cuts were evaluated for this study:

(1) Continuous miner driving a heading with blowing curtain
(2) Continuous miner driving an entry without curtain
(3) Continuous miner turning a right crosscut with blowing curtain
(4) Continuous miner turning a left crosscut with blowing curtain

Figure E-4 is a plot of shuttle car cab dust levels for four cuts (No. 1, 3, 9, and 10) when the miner was driving a heading and blowing curtain was used to ventilate the face. The x-axis corresponds to the sequential load number during the cut. For Cuts No. 1 and 3, the curtain length exceeded 100 ft, and the airflow quantities presented in Table E-1 for these cuts were measured at the face-end mouth of the curtain. For Cuts No. 9 and 10, the blowing curtain was less than 10 ft in length and served only to deflect a portion of the entry air toward the face. Curtain airflows could not be determined for Cuts No. 9 and 10, so the airflow quantities presented in Table E-3 for these cuts are the entry airflow levels. When driving a heading with the use of blowing curtain, no consistent trend was apparent in dust levels as mining depth increased. Dust exposure levels in the shuttle car cabs when loading at the face averaged 0.74 mg/m^3 during the regular-cut depth and 0.84 mg/m^3 for the deep-cut depth. This difference was not statistically significant, indicating that depth of cut had no bearing on dust levels when the miner was driving a heading and blowing curtain was used.

Figure E-5 is a plot of shuttle car cab dust levels for four cuts (No. 5, 6, 7, and 8) when the miner was driving an entry without ventilation curtain. All of the cuts without curtain occurred during the perimeter mining phase of the section development. As shown in Figure E-3, perimeter entries were driven to the left and perpendicular to satellite Entry No. S1. Intake airflow traveled down Entry No. S1, sweeping from left to right and perpendicular to the perimeter entries. During the initial sump, the scrubber exhaust was located in the crosscut between Entries No. S1 and S2, resulting in rollback into the shuttle car travel way. As the cut was advanced, the scrubber exhaust was directed into Entry No. S1 and was quickly swept away from the workings. As the miner further advanced the perimeter entry, the shuttle car cabs traveled past the pillar line separating Entries No. S1 and S2, placing them directly into intake air. A downward trend in shuttle car dust levels was evident during the development of each entry. When driving an entry without the use of blowing curtain, shuttle car cab dust levels were 77% lower (95% CIs) during the deep-cut sequence, averaging 0.30 mg/m^3 during the regular-cut depth and 0.07 mg/m^3 during the deep-cut depth. Lower dust levels during the deep-cut sequence most likely resulted from operator positioning, as described in previous sections, and may be due to improved dust capture by the scrubber. The improved scrubber capture associated with deep-cut techniques that did not use blowing curtain is discussed in the next section, "Miner-generated Dust."

Figure E-6 is a plot of shuttle car cab dust levels for Cut No. 2 when the miner was turning a right crosscut with blowing curtain. No clear trend is apparent in dust levels as mining depth increases. Dust exposures in the shuttle car cabs when loading at the face averaged 1.24 mg/m^3

during the regular-cut depth and 1.13 mg/m^3 for the deep-cut depth. This difference was not statistically significant (85% CIs), indicating that depth of cut had no bearing on dust levels during development of the right crosscut.

Figure E-7 is a plot of shuttle car cab dust levels for Cut No. 4 when the miner was turning a left crosscut with blowing curtain. Again, no clear trend in dust levels is apparent. Dust exposures in the shuttle car cabs averaged 1.49 mg/m^3 during the regular-cut depth and 1.10 mg/m^3 during the deep-cut depth. This difference was not statistically significant (85% CIs), indicating that depth of cut had no bearing on dust levels during development of the left crosscut.

The daily average dust levels for the shuttle car cabs were very low throughout the study, ranging from 0.10 to 0.42 mg/m^3, indicating that use of deep-cutting techniques did not hinder the mine's ability to limit the exposure levels of the shuttle car operators.

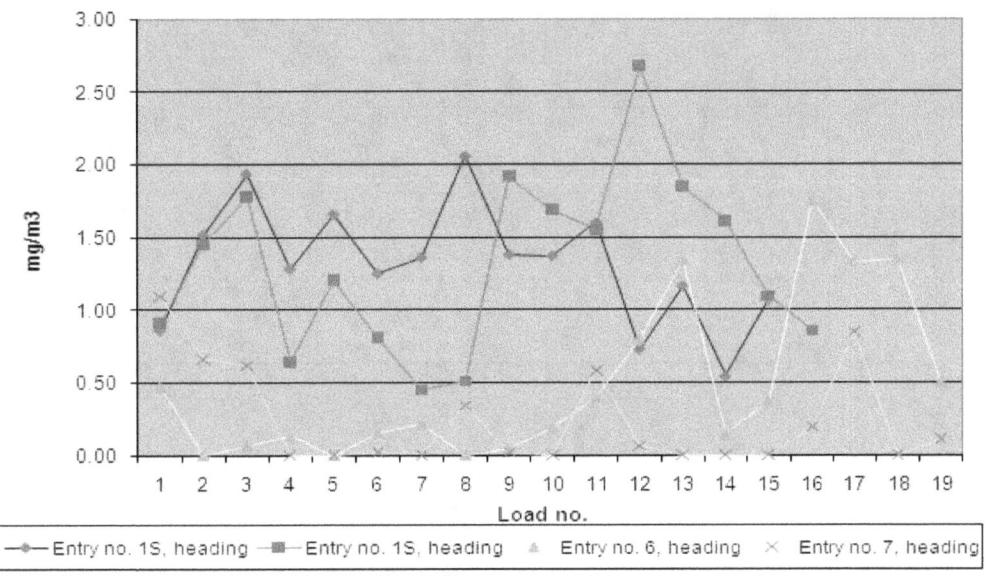

Figure E-4. Shuttle car cab dust levels at the face when driving a heading with the use of blowing curtain.

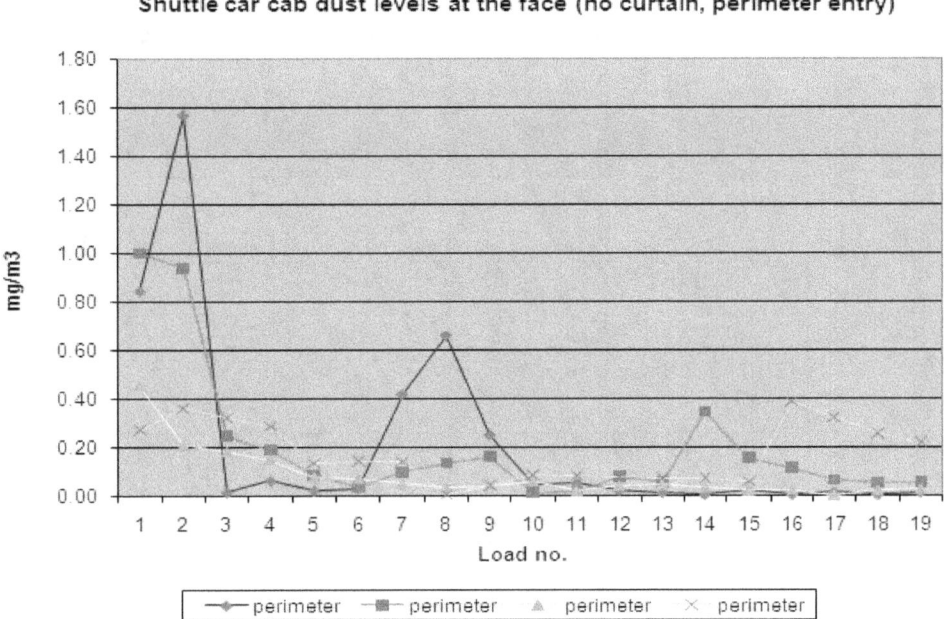

Figure E-5. Shuttle car cab dust levels at the face when driving an entry without the use of blowing curtain.

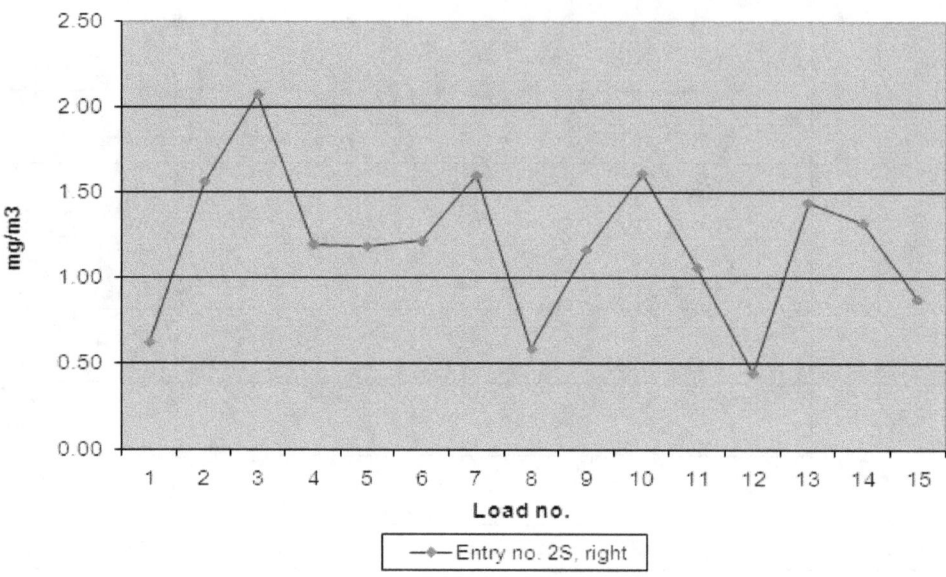

Figure E-6. Shuttle car cab dust levels at the face when turning a right crosscut with the use of blowing curtain.

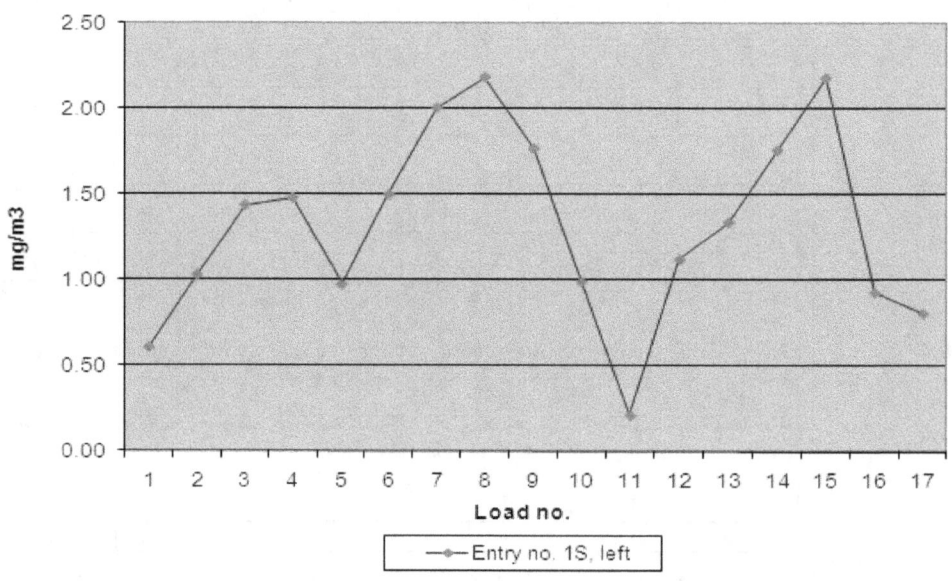

Figure E-7. Shuttle car cab dust levels at the face when turning a left crosscut with the use of blowing curtain.

Miner-generated Dust

Table E-4 shows miner-generated dust levels during the regular- and deep-cut sequences for each room development. The last column of Table E-4 adjusts dust levels based on the measured productivity for each cut sequence using an average productivity of 0.54 cars per minute.

The average miner-generated dust was lower during the deep-cut sequence when comparing it to the regular-cut sequence for each of the four cuts that did not use a ventilation curtain. In fact, miner-generated dust levels dropped 63% from an average level of 1.50 mg/m^3 during the regular-cut depth to 0.55 mg/m^3 during the deep-cut depth. Figure E-8 shows miner return dust levels at 5-sec intervals for each of the four cuts that did not use curtain. Return dust levels are clearly higher during the initial sump and slab cuts, which are visually discernable on the graph. During the second sump and slab cuts, the dust levels are much lower and cannot be differentiated on the graph. The decrease may be due to the fact that airflow reaching the face was essentially limited to what the scrubber was drawing after the miner extended beyond the influence of the ventilation curtain or primary intake airflow. The resulting lower air velocities at the face lead to improved scrubber capture efficiencies.

No statistically significant (85% CIs) difference in dust levels between the regular- and deep-cut depths was observed when blowing curtain was used to ventilate the faces. Under this scenario, when comparing the regular- to the deep-cut depth, miner-generated dust levels were quite low for both cutting depths and averaged 0.69 mg/m^3 during regular-cut depth and 0.62 mg/m^3 during deep-cut depth.

At Mine E, the 35- to 40-ft curtain setback distance was well maintained throughout each room development. This setback distance appeared to ensure fresh intake airflow for dilution, while not creating excessive air velocities at the face. Laboratory research has indicated that when the curtain-to-scrubber airflow ratio, before scrubber activation, exceeds 1.0, dust levels in the face area may increase due to decreased scrubber capture[Jayaraman et al. 1992]. It is interesting to note that Cut No. 2 at Mine E had the highest measured curtain-to-scrubber airflow ratio of 2.1 (17,800/8,500) after scrubber activation, and it also had the highest miner-generated dust levels (1.04 to 1.35 mg/m^3) for curtained cuts. Although the airflow ratio at Mine E was measured after scrubber activation, it is likely that the scrubber-off airflow ratio was still well above 1.0 because a previous laboratory study found that activation of scrubbers with airflows in the range of 6,000 to 10,000 cfm had minimal impact on blowing curtain airflows in the range of 10,000 to 14,000 cfm [Taylor et al. 1997].

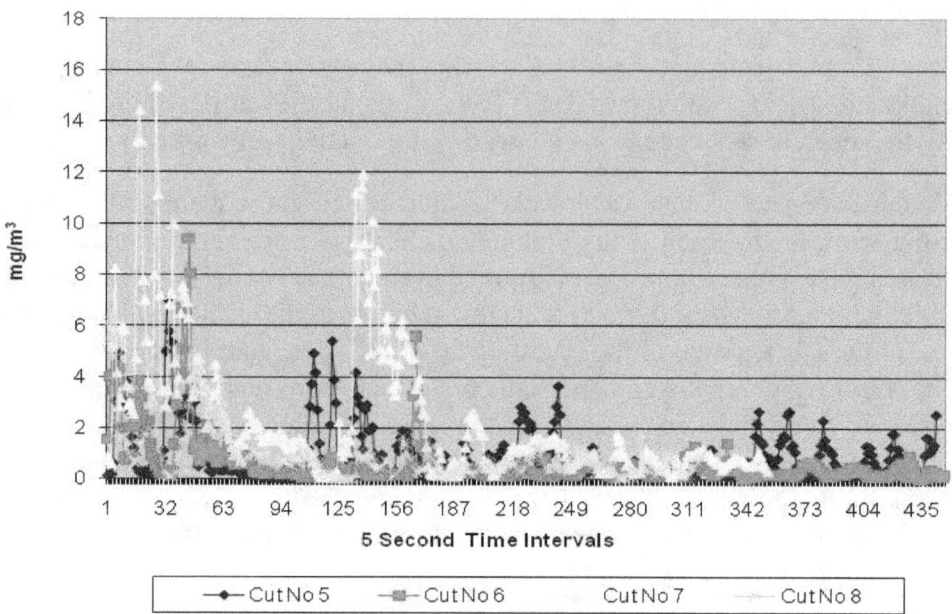

Figure E-8. Miner return dust levels at 5-sec intervals for each cut that did not use ventilation curtain.

Bolter Operations Data Analysis

Table E-5 summarizes the bolting sequences conducted during the study at Mine E. The mine used a dual-head Bucyrus RB2-88 machine to perform roof bolting operations. Routine maintenance on the bolting machine's dust collection system was performed regularly, which assured proper suction pressure for dust capture. Most bolting operations were conducted upwind of mining operations. As a result of these conditions, the bolter operators' daily average dust exposure levels were low during the study, ranging between 0.52 and 0.83 mg/m^3. Bolter-generated dust levels were also low, averaging less than 0.2 mg/m^3 on the two cuts (No. 4 and 7) where these levels could be isolated. Very few deep-cut rooms were available for analysis of bolting operations because perimeter entries were not bolted and most of the cuts on Day 2 of the study were less than 30 ft deep. Of the two that qualified as deep-cut depth with similar starting and ending conditions (Rooms No. 2 and 13), no statistically significant (85% CIs) difference in bolter operator exposures was measured when comparing bolting during the first 20 ft of depth to the remaining room depth. Room No. 4 was disqualified because bolting operations were downwind of the miner during the deep-cut depth. Use of deep-cutting techniques did not appear to adversely impact bolter operator dust exposure levels during the two cuts that were analyzed for this survey.

Dust-control Monitoring

As mentioned earlier, one of the concerns was that the scrubber would become clogged with particles as the mining cycle progressed through the deep cut, diminishing its effectiveness.

Table E-4 shows scrubber airflows at the start and end of each cut. An 8% average drop-off in scrubber airflow was observed from the start to the end of the cuts. However, this decrease in scrubber airflow did not produce a negative effect on downwind dust levels at Mine E, nor did it appear to affect the dust exposure levels of the face workers. Two factors may have contributed to this observation. First, as discussed in the "Miner-generated Dust" section of this report, dust capture appears to improve as the cutting depth increases beyond the influence of face ventilation airflow. In addition, higher pressure drops across the loaded scrubber filters may be improving scrubber efficiencies associated with respirable-sized particles to compensate somewhat for the processing of lower volumes of air.

As with the continuous miner, no degradation in the bolting machine's dust control circuit was observed during completion of the bolting sequence in the deep-cut rooms. Suction measurements were taken at the start and end of each bolting sequence and 80% of these were between 9 and 14-in Hg, which provided excellent dust control.

Summary

1. When the miner was driving a heading and blowing curtain was used to ventilate the face, dust exposures in the shuttle car cabs averaged 0.74 mg/m^3 during the regular-cut depth and 0.84 mg/m^3 during the deep-cut depth. This difference is not statistically significant (85% CIs), indicating that depth of cut had no bearing on dust levels.

2. When the miner was driving an entry without the use of blowing curtain, shuttle car cab dust levels were 77% higher during the regular-cut sequence, when compared to the deep-cut, averaging 0.30 mg/m^3 during the regular-cut depth and 0.07 mg/m^3 during the deep-cut depth. This result is significant at a 95% CIs.

3. When the miner was turning a right crosscut with blowing curtain, dust exposures in the shuttle car cabs when loading at the face averaged 1.24 mg/m^3 during the regular-cut depth and 1.13 mg/m^3 during the deep-cut depth. This difference is not statistically significant (85% CIs), indicating that depth of cut had no bearing on dust levels.

4. When the miner was turning a left crosscut with blowing curtain, dust exposures in the shuttle car cabs averaged 1.49 mg/m^3 during the regular-cut depth and 1.10 mg/m^3 during the deep-cut depth. This difference is not statistically significant (85% CIs), indicating that depth of cut had no bearing on dust levels.

5. The daily average dust levels for the shuttle car cabs were very low throughout the study, ranging from 0.10 to 0.42 mg/m^3, indicating that use of deep-cutting techniques did not hinder the mine's ability to limit the exposure levels of the shuttle car operators.

6. Miner-generated dust was 63% lower during the deep-cut sequence when comparing it to the regular-cut sequence for the four cuts that did not use ventilation curtain.

7. No statistically significant (85% CIs) difference in miner-generated dust levels was observed between the regular- and deep-cut depths when blowing curtain was used to ventilate the faces.

8. Use of deep-cutting techniques did not adversely impact bolter operator dust exposure levels during this survey.

9. An 8% average drop-off in scrubber airflow was observed from the start to the end of the cuts. However, this decrease in scrubber airflow did not produce a negative effect on downwind dust levels at Mine E nor did it affect dust exposure levels in the face area.

References Cited in Appendix E

Jayaraman NI, Jankowski RA, Whitehead KL [1992]. Optimizing continuous miner scrubbers for dust control in high coal seams. In: Proceedings of New Technology in Mine Health and Safety, SME Annual Meeting. Littleton, CO: Society for Mining, Metallurgy, and Exploration, pp. 193–205.

Taylor DT, Rider JP, Thimons ED [1997]. Impact of unbalanced intake and scrubber flows on face methane concentrations. In: Ramani RV, ed. Proceedings of the 6th International Mine Ventilation Congress. Chapter 27. Littleton, CO: Society for Mining, Metallurgy, and Exploration, pp. 169–172.

Table E-1. Shuttle car cab dust levels when loading at the face during Cuts No. 1 through 4
Note: Shaded areas show deep-cut depth, and clear areas show regular-cut depth. Airflow measured after activation of the scrubber.

Cut No.	Day	Cut Sequence	Curtain Airflow (cfm)	Face Ventilation	Car No.	Time at Face	Shuttle Car Dust (mg/m^3)	Intake Dust Level (mg/m^3)	Adjusted Shuttle Car Dust (mg/m^3)
1	1	No. 1S, Heading	6,700	Blowing	1	8:23:00 to 8:23:55	1.12	0.26	0.86
1	1	No. 1S, Heading	6,700	Blowing	2	8:26:58 to 8:27:51	1.74	0.23	1.52
1	1	No. 1S, Heading	6,700	Blowing	1	8:28:59 to 8:29:41	2.29	0.36	1.93
1	1	No. 1S, Heading	6,700	Blowing	2	8:31:25 to 8:32:32	1.65	0.37	1.28
1	1	No. 1S, Heading	6,700	Blowing	1	8:33:14 to 8:33:50	2.26	0.61	1.66
1	1	No. 1S, Heading	6,700	Blowing	2	8:36:51 to 8:37:41	1.67	0.42	1.25
1	1	No. 1S, Heading	6,700	Blowing	1	8:38:35 to 8:39:20	1.93	0.57	1.36
1	1	No. 1S, Heading	6,700	Blowing	2	8:41:16 to 8:42:01	2.56	0.50	2.05
1	1	No. 1S, Heading	6,700	Blowing	1	8:43:48 to 8:45:37	1.73	0.35	1.38
1	1	No. 1S, Heading	6,700	Blowing	3	8:51:06 to 8:52:00	1.62	0.25	1.37
1	1	No. 1S, Heading	6,700	Blowing	2	8:53:06 to 8:53:38	2.06	0.46	1.60
1	1	No. 1S, Heading	6,700	Blowing	1	8:55:54 to 8:56:35	1.01	0.28	0.73
1	1	No. 1S, Heading	6,700	Blowing	2	8:58:33 to 8:59:28	1.47	0.31	1.16
1	1	No. 1S, Heading	6,700	Blowing	1	9:01:00 to 9:01:36	1.07	0.53	0.54
1	1	No. 1S, Heading	6,700	Blowing	2	9:03:35 to 9:04:24	1.40	0.33	1.07
2	1	No. 2S, Right	17,800	Blowing	2	10:36:27 to 10:38:10	0.81	0.19	0.62
2	1	No. 2S, Right	17,800	Blowing	3	10:38:40 to 10:39:52	1.72	0.16	1.56
2	1	No. 2S, Right	17,800	Blowing	1	10:40:22 to 10:41:10	2.21	0.14	2.07
2	1	No. 2S, Right	17,800	Blowing	2	10:44:20 to 10:45:17	1.43	0.23	1.20
2	1	No. 2S, Right	17,800	Blowing	3	10:45:57 to 10:46:45	1.54	0.35	1.19
2	1	No. 2S, Right	17,800	Blowing	1	10:48:09 to 10:49:21	1.49	0.27	1.22
2	1	No. 2S, Right	17,800	Blowing	2	10:50:09 to 10:51:24	1.83	0.23	1.60
2	1	No. 2S, Right	17,800	Blowing	3	10:58:00 to 10:58:46	0.77	0.19	0.58
2	1	No. 2S, Right	17,800	Blowing	1	10:59:30 to 11:00:32	1.33	0.17	1.16
2	1	No. 2S, Right	17,800	Blowing	2	11:01:14 to 11:02:14	1.77	0.16	1.61
2	1	No. 2S, Right	17,800	Blowing	3	11:02:53 to 11:03:50	1.25	0.18	1.06
2	1	No. 2S, Right	17,800	Blowing	1	11:04:45 to 11:05:35	0.66	0.22	0.44
2	1	No. 2S, Right	17,800	Blowing	2	11:07:03 to 11:07:55	1.67	0.23	1.44
2	1	No. 2S, Right	17,800	Blowing	3	11:08:52 to 11:09:56	1.55	0.23	1.32
2	1	No. 2S, Right	17,800	Blowing	1	11:10:44 to 11:13:21	1.13	0.25	0.88
3	1	No. 1S, Heading	6,500	Blowing	3	12:23:12 to 12:24:16	1.10	0.19	0.91
3	1	No. 1S, Heading	6,500	Blowing	1	12:25:25 to 12:26:26	1.72	0.27	1.45
3	1	No. 1S, Heading	6,500	Blowing	2	12:27:34 to 12:28:25	2.12	0.35	1.77
3	1	No. 1S, Heading	6,500	Blowing	3	12:29:24 to 12:30:09	1.16	0.51	0.64
3	1	No. 1S, Heading	6,500	Blowing	1	12:31:17 to 12:32:11	1.55	0.35	1.20
3	1	No. 1S, Heading	6,500	Blowing	2	12:33:06 to 12:33:42	1.24	0.44	0.81
3	1	No. 1S, Heading	6,500	Blowing	3	12:35:14 to 12:35:53	0.70	0.25	0.45
3	1	No. 1S, Heading	6,500	Blowing	1	12:36:56 to 12:37:51	0.79	0.28	0.51
3	1	No. 1S, Heading	6,500	Blowing	2	12:38:48 to 12:39:30	2.28	0.37	1.91
3	1	No. 1S, Heading	6,500	Blowing	3	12:40:33 to 12:41:18	2.24	0.56	1.68
3	1	No. 1S, Heading	6,500	Blowing	1	12:42:46 to 12:43:35	1.81	0.26	1.55
3	1	No. 1S, Heading	6,500	Blowing	2	12:44:36 to 12:45:20	3.14	0.47	2.67
3	1	No. 1S, Heading	6,500	Blowing	3	12:46:30 to 12:47:18	2.38	0.54	1.84
3	1	No. 1S, Heading	6,500	Blowing	1	12:48:24 to 12:49:08	2.07	0.46	1.61
3	1	No. 1S, Heading	6,500	Blowing	2	12:50:48 to 12:52:08	1.29	0.20	1.09
3	1	No. 1S, Heading	6,500	Blowing	3	12:53:23 to 12:54:15	1.13	0.27	0.85
4	2	No. 1S, Left	5,400	Blowing	2	14:04:33 to 14:06:03	0.62	0.01	0.61
4	2	No. 1S, Left	5,400	Blowing	4	14:06:51 to 14:07:23	No Data	No Data	No Data
4	2	No. 1S, Left	5,400	Blowing	3	14:08:01 to 14:08:45	1.75	0.31	1.44
4	2	No. 1S, Left	5,400	Blowing	1	14:10:26 to 14:11:24	1.53	0.05	1.48
4	2	No. 1S, Left	5,400	Blowing	2	14:11:58 to 14:12:39	1.45	0.48	0.98
4	2	No. 1S, Left	5,400	Blowing	4	14:13:43 to 14:14:41	No Data	No Data	No Data
4	2	No. 1S, Left	5,400	Blowing	3	14:15:30 to 14:16:15	2.24	0.23	2.01
4	2	No. 1S, Left	5,400	Blowing	1	14:17:04 to 14:18:13	2.49	0.31	2.19
4	2	No. 1S, Left	5,400	Blowing	2	14:19:23 to 14:20:10	1.96	0.19	1.77
4	2	No. 1S, Left	5,400	Blowing	4	14:21:00 to 14:21:56	No Data	No Data	No Data
4	2	No. 1S, Left	5,400	Blowing	3	14:35:21 to 14:36:09	0.24	0.03	0.21
4	2	No. 1S, Left	5,400	Blowing	1	14:36:53 to 14:37:42	1.17	0.05	1.12
4	2	No. 1S, Left	5,400	Blowing	2	14:38:36 to 14:39:21	1.50	0.16	1.34
4	2	No. 1S, Left	5,400	Blowing	4	14:40:09 to 14:40:51	No Data	No Data	No Data
4	2	No. 1S, Left	5,400	Blowing	3	14:41:50 to 14:43:02	2.44	0.26	2.18
4	2	No. 1S, Left	5,400	Blowing	1	14:43:47 to 14:44:43	1.32	0.39	0.93
4	2	No. 1S, Left	5,400	Blowing	2	14:45:25 to 14:46:19	1.15	0.34	0.81

Table E-2. Shuttle car cab dust levels when loading at the face during Cuts No. 5 through 7
Note: Shaded areas show deep-cut depth, and clear areas show regular-cut depth.

Cut No.	Day	Cut Sequence	Curtain Airflow (cfm)	Face Ventilation	Car No.	Time at Face	Shuttle Car Dust (mg/m^3)	Intake Dust Level (mg/m^3)	Adjusted Shuttle Car Dust (mg/m^3)
5	3	Perimeter	0	No Curtain	3	8:29:07 to 8:30:15	0.86	0.02	0.84
5	3	Perimeter	0	No Curtain	2	8:31:18 to 8:32:41	1.63	0.06	1.56
5	3	Perimeter	0	No Curtain	1	8:37:45 to 8:39:12	0.03	0.01	0.01
5	3	Perimeter	0	No Curtain	3	8:39:57 to 8:40:45	0.09	0.03	0.06
5	3	Perimeter	0	No Curtain	2	8:41:50 to 8:42:51	0.04	0.02	0.02
5	3	Perimeter	0	No Curtain	1	8:43:25 to 8:44:02	0.05	0.03	0.03
5	3	Perimeter	0	No Curtain	3	8:45:49 to 8:46:43	0.45	0.03	0.42
5	3	Perimeter	0	No Curtain	2	8:47:07 to 8:47:52	0.91	0.25	0.66
5	3	Perimeter	0	No Curtain	1	8:48:21 to 8:49:05	0.32	0.07	0.25
5	3	Perimeter	0	No Curtain	3	8:50:20 to 8:51:02	0.08	0.04	0.05
5	3	Perimeter	0	No Curtain	2	8:51:59 to 8:52:43	0.10	0.05	0.05
5	3	Perimeter	0	No Curtain	1	8:53:16 to 8:54:13	0.07	0.05	0.02
5	3	Perimeter	0	No Curtain	3	8:55:14 to 8:55:52	0.07	0.06	0.01
5	3	Perimeter	0	No Curtain	2	8:57:25 to 8:58:07	0.05	0.04	0.00
5	3	Perimeter	0	No Curtain	1	8:58:44 to 8:59:21	0.06	0.04	0.02
5	3	Perimeter	0	No Curtain	3	9:00:27 to 9:01:17	0.05	0.05	0.00
5	3	Perimeter	0	No Curtain	2	9:02:16 to 9:03:00	0.05	0.04	0.01
5	3	Perimeter	0	No Curtain	1	9:03:29 to 9:04:05	0.04	0.04	0.00
5	3	Perimeter	0	No Curtain	3	9:05:10 to 9:05:55	0.06	0.05	0.01
6	3	Perimeter	0	No Curtain	2	10:13:52 to 10:14:50	1.03	0.03	1.00
6	3	Perimeter	0	No Curtain	1	10:15:06 to 10:15:32	1.01	0.07	0.94
6	3	Perimeter	0	No Curtain	3	10:16:52 to 10:17:31	0.37	0.13	0.25
6	3	Perimeter	0	No Curtain	2	10:18:28 to 10:19:07	0.32	0.13	0.19
6	3	Perimeter	0	No Curtain	1	10:19:30 to 10:20:07	0.21	0.12	0.08
6	3	Perimeter	0	No Curtain	3	10:21:20 to 10:22:04	0.13	0.09	0.04
6	3	Perimeter	0	No Curtain	2	10:23:08 to 10:24:00	0.19	0.10	0.10
6	3	Perimeter	0	No Curtain	1	10:24:30 to 10:25:06	0.21	0.08	0.13
6	3	Perimeter	0	No Curtain	3	10:27:14 to 10:28:08	0.26	0.10	0.16
6	3	Perimeter	0	No Curtain	1	10:29:29 to 10:30:11	0.09	0.08	0.01
6	3	Perimeter	0	No Curtain	3	10:31:49 to 10:32:38	0.07	0.05	0.02
6	3	Perimeter	0	No Curtain	1	10:34:49 to 10:35:25	0.13	0.06	0.08
6	3	Perimeter	0	No Curtain	3	10:36:30 to 10:37:12	0.14	0.09	0.05
6	3	Perimeter	0	No Curtain	1	10:39:23 to 10:39:53	0.42	0.07	0.34
6	3	Perimeter	0	No Curtain	3	10:40:52 to 10:41:28	0.18	0.03	0.15
6	3	Perimeter	0	No Curtain	4	10:44:20 to 10:45:18	N/A	N/A	N/A
6	3	Perimeter	0	No Curtain	3	10:45:47 to 10:46:30	0.11	0.05	0.06
6	3	Perimeter	0	No Curtain	4	10:48:36 to 10:49:20	N/A	N/A	N/A
6	3	Perimeter	0	No Curtain	3	10:50:20 to 10:51:00	0.22	0.17	0.05
7	3	Perimeter	0	No Curtain	1	11:03:20 to 11:04:02	0.47	0.02	0.45
7	3	Perimeter	0	No Curtain	3	11:04:53 to 11:05:33	0.22	0.01	0.21
7	3	Perimeter	0	No Curtain	4	11:05:55 to 11:06:40	N/A	N/A	N/A
7	3	Perimeter	0	No Curtain	1	11:07:40 to 11:08:12	0.16	0.01	0.15
7	3	Perimeter	0	No Curtain	3	11:09:19 to 11:09:57	0.10	0.02	0.08
7	3	Perimeter	0	No Curtain	4	11:10:23 to 11:11:02	N/A	N/A	N/A
7	3	Perimeter	0	No Curtain	1	11:11:25 to 11:12:00	0.07	0.02	0.05
7	3	Perimeter	0	No Curtain	3	11:13:34 to 11:14:38	0.05	0.02	0.03
7	3	Perimeter	0	No Curtain	4	11:15:05 to 11:15:46	N/A	N/A	N/A
7	3	Perimeter	0	No Curtain	1	11:16:08 to 11:16:48	0.08	0.02	0.06
7	3	Perimeter	0	No Curtain	3	11:18:59 to 11:19:37	0.04	0.02	0.02
7	3	Perimeter	0	No Curtain	4	11:21:11 to 11:22:00	N/A	N/A	N/A
7	3	Perimeter	0	No Curtain	1	11:22:32 to 11:23:19	0.06	0.02	0.05
7	3	Perimeter	0	No Curtain	3	11:23:55 to 11:24:36	0.05	0.02	0.03
7	3	Perimeter	0	No Curtain	4	11:25:24 to 11:26:14	N/A	N/A	N/A
7	3	Perimeter	0	No Curtain	1	11:26:45 to 11:27:33	0.05	0.03	0.02
7	3	Perimeter	0	No Curtain	3	11:28:53 to 11:29:42	0.05	0.06	0.00
7	3	Perimeter	0	No Curtain	4	11:30:36 to 11:31:13	N/A	N/A	N/A
7	3	Perimeter	0	No Curtain	1	11:31:42 to 11:32:40	0.06	0.04	0.03

Table E-3. Shuttle car cab dust levels when loading at the face during Cuts No. 8 through 10

Note: Shaded areas show deep-cut depth, and clear areas show regular-cut depth.

Cut No.	Day	Cut Sequence	Curtain Airflow (cfm)	Face Ventilation	Car No.	Time at Face	Shuttle Car Dust (mg/m^3)	Intake Dust Level (mg/m^3)	Adjusted Shuttle Car Dust (mg/m^3)
8	3	Perimeter	0	No Curtain	1	11:43:45 to 11:44:48	0.28	0.00	0.27
8	3	Perimeter	0	No Curtain	3	11:45:52 to 11:46:40	0.36	0.00	0.36
8	3	Perimeter	0	No Curtain	4	11:48:00 to 11:48:49	N/A	N/A	N/A
8	3	Perimeter	0	No Curtain	1	11:49:26 to 11:50:03	0.30	0.01	0.29
8	3	Perimeter	0	No Curtain	3	11:50:36 to 11:51:26	0.15	0.01	0.13
8	3	Perimeter	0	No Curtain	4	11:52:01 to 11:52:42	N/A	N/A	N/A
8	3	Perimeter	0	No Curtain	1	11:53:05 to 11:53:46	0.15	0.02	0.14
8	3	Perimeter	0	No Curtain	3	11:54:01 to 11:54:48	0.02	0.02	0.00
8	3	Perimeter	0	No Curtain	4	12:05:00 to 12:06:27	N/A	N/A	N/A
8	3	Perimeter	0	No Curtain	1	12:06:55 to 12:07:34	0.09	0.00	0.08
8	3	Perimeter	0	No Curtain	4	12:28:09 to 12:28:51	N/A	N/A	N/A
8	3	Perimeter	0	No Curtain	1	12:29:17 to 12:29:52	0.08	0.00	0.08
8	3	Perimeter	0	No Curtain	3	12:30:50 to 12:32:29	0.08	0.01	0.07
8	3	Perimeter	0	No Curtain	4	12:32:52 to 12:33:34	N/A	N/A	N/A
8	3	Perimeter	0	No Curtain	1	12:34:10 to 12:34:44	0.07	0.02	0.06
8	3	Perimeter	0	No Curtain	3	12:35:40 to 12:36:30	0.40	0.01	0.39
8	3	Perimeter	0	No Curtain	4	12:37:00 to 12:37:38	N/A	N/A	N/A
8	3	Perimeter	0	No Curtain	1	12:38:55 to 12:40:35	0.27	0.02	0.25
8	3	Perimeter	0	No Curtain	3	12:41:12 to 12:41:53	0.24	0.02	0.22
9	3	No. 6, Heading	0	No Curtain	2	13:19:52 to 13:20:28	0.47	0.00	0.47
9	3	No. 6, Heading	0	No Curtain	1	13:21:17 to 13:21:56	0.10	0.14	0.00
9	3	No. 6, Heading	0	No Curtain	3	13:22:26 to 13:23:02	0.10	0.03	0.06
9	3	No. 6, Heading	0	No Curtain	2	13:23:33 to 13:24:08	0.17	0.04	0.13
9	3	No. 6, Heading	0	No Curtain	1	13:26:07 to 13:26:51	0.38	0.43	0.00
9	3	No. 6, Heading	0	No Curtain	3	13:27:44 to 13:28:34	0.18	0.03	0.15
9	3	No. 6, Heading	0	No Curtain	2	13:29:08 to 13:29:50	0.25	0.03	0.21
9	3	No. 6, Heading	<14,400	Blowing	1	13:31:27 to 13:32:00	0.06	0.07	0.00
9	3	No. 6, Heading	<14,400	Blowing	3	13:33:32 to 13:34:06	0.07	0.03	0.04
9	3	No. 6, Heading	<14,400	Blowing	2	13:34:33 to 13:35:14	0.19	0.01	0.18
9	3	No. 6, Heading	<14,400	Blowing	1	13:36:00 to 13:36:43	0.44	0.05	0.40
9	3	No. 6, Heading	<14,400	Blowing	3	13:37:08 to 13:37:48	0.86	0.06	0.80
9	3	No. 6, Heading	<14,400	Blowing	2	13:38:12 to 13:38:49	1.42	0.08	1.34
9	3	No. 6, Heading	<14,400	Blowing	1	13:40:22 to 13:41:11	0.22	0.07	0.14
9	3	No. 6, Heading	<14,400	Blowing	3	13:41:40 to 13:42:15	0.40	0.02	0.37
9	3	No. 6, Heading	<14,400	Blowing	2	13:42:37 to 13:43:16	1.84	0.09	1.75
9	3	No. 6, Heading	<14,400	Blowing	1	13:44:16 to 13:44:55	1.39	0.05	1.33
9	3	No. 6, Heading	<14,400	Blowing	3	13:45:23 to 13:45:45	1.42	0.07	1.35
9	3	No. 6, Heading	<14,400	Blowing	2	13:47:04 ro 13:47:57	0.57	0.07	0.50
10	3	No. 7, Heading	0	No Curtain	2	14:05:39 to 14:06:17	1.14	0.05	1.09
10	3	No. 7, Heading	0	No Curtain	1	14:07:06 to 14:08:00	0.97	0.31	0.66
10	3	No. 7, Heading	0	No Curtain	3	14:08:41 to 14:09:19	0.90	0.28	0.62
10	3	No. 7, Heading	0	No Curtain	2	14:10:16 to 14:10:56	0.15	0.22	0.00
10	3	No. 7, Heading	0	No Curtain	1	14:11:44 to 14:12:24	0.08	0.08	0.00
10	3	No. 7, Heading	0	No Curtain	3	14:12:54 to 14:13:33	0.12	0.10	0.02
10	3	No. 7, Heading	0	No Curtain	2	14:13:59 to 14:14:47	0.10	0.10	0.00
10	3	No. 7, Heading	0	No Curtain	1	14:15:48 to 14:16:26	0.45	0.11	0.34
10	3	No. 7, Heading	0	No Curtain	3	14:16:55 to 14:17:38	0.12	0.10	0.02
10	3	No. 7, Heading	0	No Curtain	2	14:18:17 to 14:19:00	0.07	0.09	0.00
10	3	No. 7, Heading	0	No Curtain	1	14:19:23 to 14:20:01	0.66	0.08	0.58
10	3	No. 7, Heading	0	No Curtain	3	14:21:05 to 14:21:47	0.17	0.10	0.06
10	3	No. 7, Heading	0	No Curtain	2	14:22:56 to 14:23:36	0.09	0.11	0.00
10	3	No. 7, Heading	0	No Curtain	1	14:24:04 to 14:25:02	0.10	0.11	0.00
10	3	No. 7, Heading	<24,300	Blowing	3	14:26:01 to 14:26:44	0.07	0.12	0.00
10	3	No. 7, Heading	<24,300	Blowing	2	14:27:05 to 14:27:43	0.28	0.09	0.19
10	3	No. 7, Heading	<24,300	Blowing	1	14:28:03 to 14:28:44	1.03	0.19	0.84
10	3	No. 7, Heading	<24,300	Blowing	3	14:30:09 to 14:30:54	0.12	0.15	0.00
10	3	No. 7, Heading	<24,300	Blowing	2	14:31:18 to 14:32:07	0.20	0.09	0.11

Table E-4. Miner-generated dust levels during the deep- and regular-cut depths
Note: Dust levels are adjusted for productivity.

Cut No.	Cut Sequence	Depth	Curtain Airflow (cfm)	Starting Scrubber Airflow (cfm)	Ending Scrubber Airflow (cfm)	Miner Intake Dust Level (mg/m³)	Miner Return Dust Level (mg/m³)	Miner Generated (mg/m³)	Cars Per Minute	Miner Generated Adjusted (mg/m³)
1	No. 1S, Heading	Regular	6,700	8,500	N/A	0.34	0.39	0.06	0.40	0.08
1	No. 1S, Heading	Deep	6,700	N/A	5,500	0.29	0.23	0.00	0.45	0.00
2	No. 2S, Right	Regular	17,800	8,500	N/A	0.21	1.14	0.93	0.37	1.35
2	No. 2S, Right	Deep	17,800	N/A	8,500	0.22	1.18	0.96	0.50	1.04
3	No. 1S, Heading	Regular	6,500	8,500	N/A	0.32	0.79	0.48	0.55	0.47
3	No. 1S, Heading	Deep	6,500	N/A	7,800	0.33	1.12	0.79	0.51	0.83
4	No. 1S, Left	Regular	5,400	8,800	N/A	0.21	0.87	0.66	0.58	0.61
4	No. 1S, Left	Deep	5,400	N/A	8,300	0.16	0.45	0.29	0.32	0.48
5	Perimeter	Regular	0	7,800	N/A	0.04	1.11	1.07	0.45	1.28
5	Perimeter	Deep	0	N/A	5,500	0.04	0.61	0.56	0.64	0.47
6	Perimeter	Regular	0	6,800	N/A	0.10	1.09	0.99	0.63	0.85
6	Perimeter	Deep	0	N/A	7,100	0.07	0.31	0.24	0.46	0.28
7	Perimeter	Regular	0	7,100	N/A	0.02	3.68	3.66	0.72	2.75
7	Perimeter	Deep	0	N/A	7,100	0.03	1.14	1.11	0.60	1.00
8	Perimeter	Regular	0	7,800	N/A	0.01	0.85	0.85	0.40	1.14
8	Perimeter	Deep	0	N/A	7,800	0.01	0.24	0.23	0.29	0.43
9	No. 6, Heading	Regular	<14,400	7,800	N/A	0.07	0.80	0.73	0.63	0.62
9	No. 6, Heading	Deep	<14,400	N/A	7,500	0.06	1.18	1.12	0.75	0.81
10	No. 7, Heading	Regular	<24,300	7,500	N/A	0.16	1.54	1.38	0.75	0.99
10	No. 7, Heading	Deep	<24,300	N/A	7,500	0.13	0.86	0.74	0.72	0.55

Table E-5. Bolter-generated dust levels and operator exposures for each bolting sequence

Room No.	Entry	Depth	Position With Respect To Miner	Curtain Airflow (cfm)	Face Ventilation	Bolter Suction Left/Right (in Hg)	Left-Side Bolter Dust Levels (mg/m³)	Right-Side Bolter Dust Levels (mg/m³)	Upwind Bolter Dust Levels (mg/m³)	Downwind Bolter Dust Levels (mg/m³)	Bolter Generated Dust (mg/m³)
1	No. 3S, Left	Regular	Downwind	0	No Curtain	10/10	0.37	0.31	N/A	N/A	N/A
2	No. 1S, Heading	Regular	Upwind	0	No Curtain	10/11	1.06	1.01	N/A	N/A	N/A
2	No. 1S, Heading	Deep	Upwind	0	No Curtain	10/11	0.89	0.60	N/A	N/A	N/A
3	No. 1S, Right	Regular	Upwind	0	No Curtain	10/10	0.30	0.52	N/A	N/A	N/A
4	No. 2S, Right	Regular	Upwind	11,700	Blowing	9/9	0.20	0.08	0.21	0.62	0.41
4	No. 2S, Right	Deep	Downwind	11,700	Blowing	9/9	0.58	0.56	0.51	0.36	0.00
5	No. 3S, Heading	Regular	Downwind	0	No Curtain	5/10	1.29	1.33	N/A	N/A	N/A
6	No. 2S, Heading	Regular	Upwind	0	No Curtain	14/10	0.63	0.79	N/A	N/A	N/A
7	No. 2S, Right	Regular	Upwind	4,400	Blowing	12/9	0.43	0.08	0.10	0.23	0.13
8	No. 3S, Heading	Regular	Upwind	0	No Curtain	12/9	0.75	0.53	N/A	N/A	N/A
9	No. 1S, Heading	Regular	Upwind	0	No Curtain	12/7	0.38	0.64	N/A	N/A	N/A
10	No. 2S, Left	Regular	Downwind	0	No Curtain	12/7	0.86	0.94	N/A	N/A	N/A
11	No. 1S, Left	Regular	Upwind	0	No Curtain	12/7	0.42	0.19	N/A	N/A	N/A
12	No. 3S, Left	Regular	Downwind	0	No Curtain	11/8	0.19	0.43	N/A	N/A	N/A
13	No. 6, Heading	Regular	Upwind	0	No Curtain	11/8	0.00	0.40	N/A	N/A	N/A
13	No. 6, Heading	Deep	Upwind	0	No Curtain	11/8	0.28	0.88	N/A	N/A	N/A

Appendix F: MINE F CASE STUDY

Mine-specific Information

Mine F used two Joy 14CM15 continuous miners. The miner spray configuration is shown in Figure F-1. A total of 52 dust suppression sprays were available to be operated at a pressure of not less than 60 psi. Forty-six were operated, but the mine ventilation plan required that only 34 of the sprays be operational. The mine ventilation plan required the line curtain to be maintained within 10 ft outby the continuous miner boom when exhausting ventilation was used. If blowing ventilation was used, the curtain was required to be maintained within 10 ft of the deepest point of penetration at the beginning of the mining cycle; as mining progresses the curtain was to be extended to the last row of permanent supports. The plan also specified minimum curtain airflows to be as follows: when the scrubber is being used, the minimum face curtain airflow for exhausting face ventilation system will be 6000 cfm or the volume of air necessary for a mean entry air velocity of 60 fpm, whichever is greater. For a blowing face ventilation system, the quantity of air discharged from the inby end of the curtain will be balanced with the rated volume of air discharged by the scrubber. Otherwise, without the scrubber, a minimum curtain airflow of 5000 cfm will be maintained.

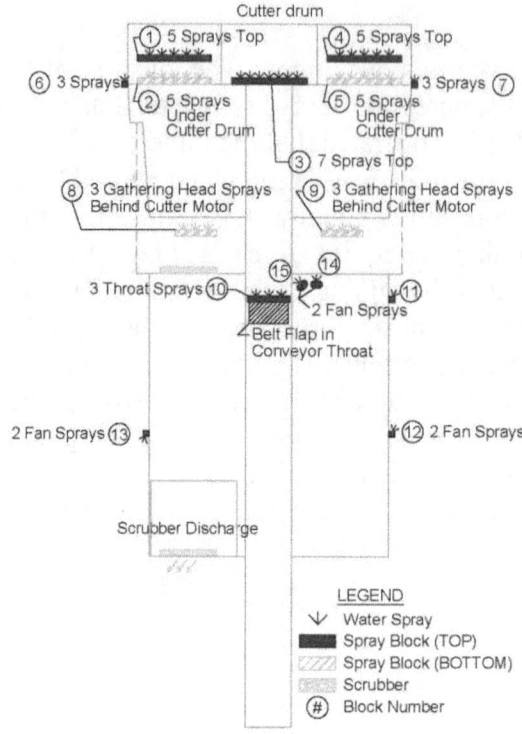

Figure F-1. Mine F continuous miner spray configuration.

Figure F-2 shows a plan view of the continuous mining section studied at Mine F. When cutting on the right side of the section (Entries No. 5, 6, and 7), the mine used blowing curtain ventilation. Exhausting curtain was used when cutting on the left side (Entries No. 1, 2, and 3). Entry No. 4, which could be ventilated with either curtain arrangement, used blowing curtain during the study. Bolting operations were ventilated similar to mining operations. The active section was ventilated using a sweeping configuration from right to left. Entry No. 7 served as the intake, and Entry No. 1 served as the return. Entries No. 2 through 6 were neutral. The belt was in Entry No. 4. The main intake airflow measured in the last open crosscut ranged from 29,800 to 65,800 cfm. Mining height varied from 6 to 6.5 ft, with the miner extracting approximately 1.5 ft of top-rock to reach a competent roof strata. Entry widths were 20 ft. Deep-cut depths at Mine F were 30 ft, and the mine used a cutting sequence consisting of the following:

(1) 20-ft sump cut, right side
(2) 20-ft slab cut, left side
(3) 10-ft sump cut, left side
(4) 10-ft slab cut, right side

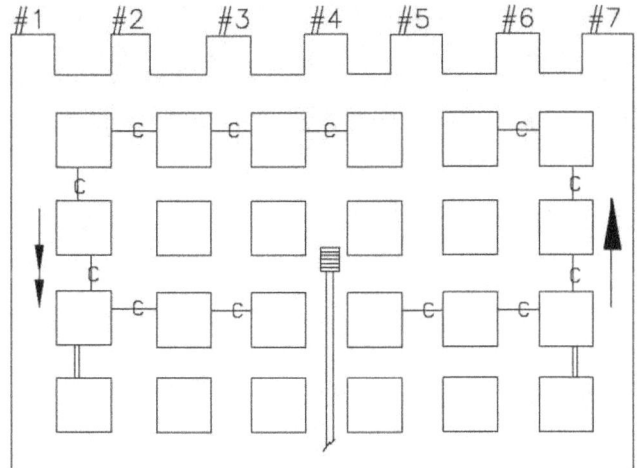

Figure F-2. Plan view of the operating section at Mine F.

The Mine F dust survey was conducted over 3 consecutive days from January 12 through 14, 2010. An equipment preparation error resulted in the loss of PDM data collected during Day 1 of the study.

Shuttle Car Data Analysis

Tables F-1, F-2, and F-3 show the shuttle car data collected during the study. One shuttle car load was determined to be invalid on Day 2 (Table F-2) of the study because the miner was down for several minutes while the car was located in the face area. When driving Entry No. 4 on Day 3 (Table F-3), the mine used the third shuttle car, which was dedicated for use on the left side of the section. The third car was not equipped with an area sampling package, so data were not collected for half of the loads during this cut. The right-side Joy 14CM15 continuous miner and two Highlander 1021 9-ton shuttle cars were evaluated for each cut during the study. All shuttle cars had a standard cab configuration, placing the operators on the curtain side of the entry. A total of six shuttle car loads were needed to advance the face 10 ft; therefore, the last six loads in the cutting sequence occurred during deep-cut depth, including a 10-ft sump cut on the left side of the entry and a 10-ft slab cut on the right side. Due to the narrowness of pillars (45 x 60 ft) at Mine F, deep-cuts were not used to drive crosscut entries. Therefore; all collected deep-cut data are for heading developments.

Figure F-3 is a plot of shuttle car dust exposures for each of the eights cuts. The x-axis corresponds to the sequential load number during the cut. The number of loads required to complete each cut varied from 15 to 19 due to factors including shuttle car loading, ceiling height, and origin of the cut with respect to the last row of roof bolts. The average number of loads to complete a 30-ft cut was 17.5; therefore, the transition between regular-cut depth and deep-cut depth occurred between Loads No. 11 and 12, depicted by a vertical line in Figure F-3. No clear pattern is visible when comparing dust levels on the right side (deep-cut depth) of the transition to levels on the left side (regular-cut depth). When cutting a heading and using blowing face ventilation, shuttle car dust levels at the face averaged 1.77 mg/m^3 during the regular-cut sequence and 2.13 mg/m^3 during the deep-cut sequence. This difference was not statistically significant (85% CIs), indicating that the use of deep-cutting techniques did not affect shuttle car dust levels at Mine F.

The daily average dust levels for the shuttle car cabs were low throughout the study, ranging from 0.21 to 0.61 mg/m^3. Shuttle car operator exposures benefited from large amounts (> 60,000 cfm) of primary airflow for dilution of miner-generated dust in the travel ways and from the mine's practice of wetting the travel ways to reduce dust entrainment from vehicle tires.

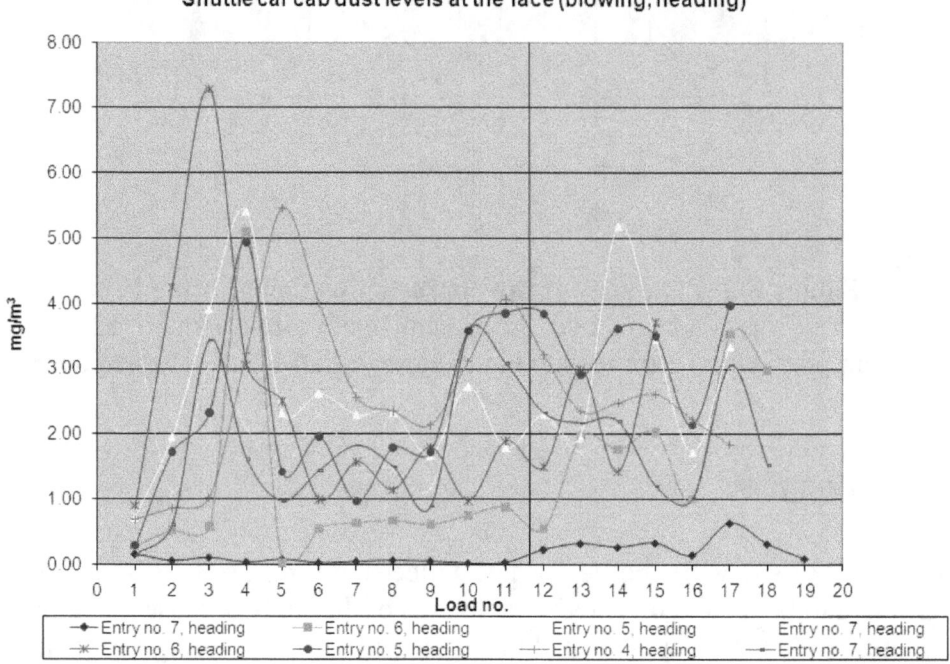

Figure F-3. Shuttle car cab dust levels when loading at the active face.

Miner-generated Dust

Table F-4 shows miner-generated dust levels during the regular- and deep-cut sequences for each room development. The last column of Table F-4 adjusts dust levels based on the measured productivity using the previously described normalization process and Mine F's average cars loaded per minute of 0.43.

Miner-generated dust levels were very low during both the regular- and the deep-cut sequences, averaging 0.78 mg/m^3 for the regular-cut depth and 0.69 mg/m^3 for the deep-cut depth. The differences between these dust levels was not statistically significant (85% CIs), indicating that using deep-cutting techniques did not impact miner dust generation. Low miner-generated dust levels were achieved through good scrubber maintenance and high section airflow (> 60,000 cfm), which effectively diluted dust not captured by the scrubber.

Miner Operator Data Analysis

Table F-5 shows miner operator dust exposure levels as measured with a PDM device for each cut. As mentioned previously, the PDM data collected on Day 1 of the study were lost due to an equipment preparation error. The levels presented in the table isolate production time only and are not indicative of shift average levels, which were much lower. Dust exposures are adjusted for productivity. Depth of cut did not significantly (85% CIs) impact the miner operator's exposures, which averaged 2.63 mg/m^3 during the regular-cut sequence and 2.55 mg/m^3 during the deep-cut sequence.

During the first 2 days of our study at Mine F, blowing brattice curtain was hung from the third row of bolts outby the face at the start of each cut. The curtain was advanced to the last row of bolts after the face was advanced 10 ft. This practice did not allow the miner operator to stand at the mouth of the curtain, where dust levels were much lower. On Day 3, greater curtain setback distances of up to 40 ft were observed and the operator was able to maintain a position at the mouth of the curtain throughout the cuts. As a result of this improved positioning, the miner operator's dust exposure levels during production decreased from an average of 4.17 mg/m^3 on Day 2 to 0.22 mg/m^3 on Day 3.

The miner operator's average dust exposures over the entire sampling period were 1.71 mg/m^3 on Day 2 and 0.81 mg/m^3 on Day 3. These levels include downtime, cuts mined on the left side (exhausting curtain side) of the section, and cuts mined on the right side (blowing curtain side) of the section. Each side of the section had its own continuous mining machine and roof bolter. The miner operator alternated between taking cuts on the left and right side of the section throughout the study. This practice allowed maintenance to be conducted on the idled mining machine without interrupting production. Routine miner maintenance included cleaning the scrubber screen, as well as replacing worn bits and clogged water sprays before each cut.

Bolter Operations Data Analysis

Bolting operations at Mine F were completed using a Fletcher RRII-13BC-F bolter. Table F-6 shows bolter-generated dust levels as well as exposure levels for the intake- and return-side bolter operators at regular- and deep-cut depths. Bolter dust generation data for Room No. 2 were invalid because the downwind area sampling package could not be placed in the immediate airflow return for the machine due to its position in the heading. Like the miner faces, the bolting faces were ventilated with blowing curtain.

No significant (85% CIs) difference was observed when comparing the average dust exposure of the intake-side bolter operator (1.14 mg/m^3) to the return-side operator (1.35 mg/m^3) during bolting operations. Likewise, bolter operator exposure levels were not statistically significant (85% CIs) when comparing bolting operations at regular-cut depth to deep-cut depth, averaging 0.87 mg/m^3 during regular-cut depth and 1.62 mg/m^3 during deep-cut depth. While not statistically significant, dust levels during the deep-cut portion of the room were double those during the regular-cut, which may have resulted from the fact that the ventilation curtain was not advanced as the bolter moved toward the face. Unlike the continuous miner, the bolting machine does not have a secondary air-moving device, such as a scrubber or water fan spray system, to maintain the flow of air toward the face as the machine advances away from the mouth of the curtain. Therefore, ventilation air tends to short circuit behind the bolter following the path of least resistance as the machine progresses beyond the influence of the blowing curtain. This occurrence progressively diminishes dust dilution at the operator positions. Data suggest that periodically advancing the curtain during the bolting cycle at Mine F may reduce bolter operator dust exposure levels. The mine did an excellent job at keeping the bolting machine upwind of continuous miner operations. As measured over the entire sampling period, the bolter operators' average dust exposure levels were low during the study, ranging from 0.61 mg/m^3 to 1.02 mg/m^3.

No statistically significant (85% CIs) difference was observed in dust levels generated by the bolting machine when comparing regular-cut depth to deep-cut depth. Bolter-generated dust levels averaged 0.35 mg/m^3 during the study at an average curtain airflow rate of 5,500 cfm.

Dust-control Monitoring

Throughout the study, the curtain lines that provided ventilation to the active faces were well maintained. Blowing curtain was supported by pogo sticks to maintain a consistent quantity of airflow during each cut. Before commencing mining at the active face, an attempt was made to match the curtain airflow to the rated scrubber airflow by adjusting the curtain overlap into the open crosscut. Table F-4 shows curtain and scrubber airflow measurements for each cut. Curtain airflow measurements were made after activation of the miner scrubber. The average scrubber airflow dropped 5% from 8,860 cfm at the start of the cut to 8,440 cfm at completion due to material loading on the 20-mesh scrubber screen. The average blowing curtain airflow quantity was 6,650 cfm, resulting in a curtain-to-scrubber airflow ratio of approximately 0.8 with the scrubber running. This ratio is well below ideal according to past research that found the best ratio to be 1.0 before activation of the scrubber [Jayaraman et al. 1992]. The identified ideal ratio involved a configuration with the scrubber inlet located under the boom of the miner. The scrubber inlet on the miner used at Mine F was located on the top left side of the machine just behind the boom hinge point. It is unclear whether the recommended operating ratio of 1.0 would apply to this type of configuration. The mine could benefit from determining the ideal curtain-to-scrubber airflow ratio to minimize dust levels in working areas around the continuous miner using its pDR dust sampling device.

The largest degradation in scrubber airflow occurred during Cut No. 8, when scrubber airflow dropped 10% from 9,100 cfm at the start of the cut to 8,200 cfm at the end. Despite this drop, no degradation in scrubber dust capture was measured during the cut. When comparing regular- to deep-cut depth during Cut No. 8, miner-generated dust levels were 0.58 mg/m^3 during regular-cut depth and 0.29 mg/m^3 during deep-cut depth. Likewise, the miner and shuttle car operators' dust exposures did not worsen during the deep-cut sequence. For this study, the scrubber screen was tapped and back-flushed before each cut. This is a good practice to assure proper scrubber function.

As with the continuous miner, no degradation in the bolting machine's dust control circuit was observed during completion of the bolting sequence in the deep-cut rooms. Suction measurements were taken periodically throughout the bolting sequences and all measurements were between 12 and 15-in Hg, which appeared to provide good dust control.

Summary

1. When using blowing curtain, shuttle car dust levels at the face averaged 1.77 mg/m^3 during the regular-cut sequence and 2.13 mg/m^3 during the deep-cut sequence. This

difference was not statistically significant (85% CIs), indicating that the use of deep-cuts did not affect shuttle car dust levels at Mine F.

2. The daily average dust levels for the shuttle car cabs were low throughout the study, ranging from 0.21 to 0.61 mg/m^3. Shuttle car operator exposures benefitted from large amounts (> 60,000 cfm) of primary airflow for dilution of miner-generated dust in the travel ways and from the mine's practice of wetting the travel ways to reduce dust entrainment from vehicle tires.

3. Miner-generated dust levels were very low during both cutting sequences, averaging 0.78 mg/m^3 during the regular-cut sequence and 0.69 mg/m^3 during the deep-cut sequence. The differences between these dust levels were not statistically significant (85% CIs), indicating that using deep-cutting techniques did not impact miner dust generation.

4. Isolating for production time only, depth of cut did not significantly (85% CIs) impact the miner operator's exposures, which averaged 2.63 mg/m^3 during the regular-cut sequence and 2.55 mg/m^3 the deep-cut sequence.

5. When curtain setback distances allowed miner operators to stand at the mouth of the blowing curtain during the entire cut, their dust exposure levels during production were dramatically reduced from 4.17 to 0.22 mg/m^3.

6. The miner operator's average dust exposure levels over the entire sampling period were 1.71 mg/m^3 on Day 2 and 0.81 mg/m^3 on Day 3.

7. Bolter operator exposure levels were not statistically significant (85% CIs) when comparing bolting operations at regular-cut depth to deep-cut depth, averaging 0.87 mg/m^3 during regular-cut depth and 1.62 mg/m^3 during deep-cut depth.

8. As measured over the entire sampling period, the bolter operators' average dust exposure levels were low during the study, ranging from 0.61 to 1.02 mg/m^3.

9. No statistically significant (85% CIs) difference was observed in dust levels generated by the bolting machine when comparing regular-cut depth to deep-cut depth. Bolter-generated levels averaged 0.35 mg/m^3 during the study at an average curtain airflow rate of 5,500 cfm.

10. The mine could benefit from using its pDR dust sampling device to determine the ideal curtain-to-scrubber airflow ratio to minimize dust levels in working areas around the mining machine.

Reference Cited in Appendix F

Jayaraman NI, Jankowski RA, Whitehead KL [1992]. Optimizing continuous miner scrubbers for dust control in high coal seams. In: Proceedings of New Technology in Mine Health and Safety, SME Annual Meeting. Littleton, CO: Society for Mining, Metallurgy, and Exploration, pp. 193–205.

Table F-1. Shuttle car cab dust levels when loading at the face on Day 1
Note: Shaded areas show deep-cut depth, and clear areas show regular-cut depth.

Cut No.	Day	Cut Sequence	Curtain Airflow	Face Ventilation	Car No.	Time at Face	Shuttle Car Dust (mg/m^3)	Intake Dust Level (mg/m^3)	Adjusted Shuttle Car Dust (mg/m^3)
1	1	No. 7, heading	6,500	Blowing	2	9:40:36 to 9:42:00	0.17	0.01	0.16
1	1	No. 7, heading	6,500	Blowing	1	9:42:46 to 9:43:54	0.07	0.01	0.06
1	1	No. 7, heading	6,500	Blowing	2	9:44:54 to 9:46:34	0.12	0.01	0.11
1	1	No. 7, heading	6,500	Blowing	1	9:47:17 to 9:49:03	0.04	0.01	0.03
1	1	No. 7, heading	6,500	Blowing	2	9:49:40 to 9:52:40	0.09	0.01	0.08
1	1	No. 7, heading	6,500	Blowing	1	9:53:22 to 9:54:40	0.03	0.01	0.02
1	1	No. 7, heading	6,500	Blowing	2	9:55:28 to 9:56:48	0.06	0.01	0.05
1	1	No. 7, heading	6,500	Blowing	1	9:57:29 to 9:59:04	0.07	0.01	0.06
1	1	No. 7, heading	6,500	Blowing	2	10:00:04 to 10:02:47	0.06	0.01	0.05
1	1	No. 7, heading	6,500	Blowing	1	10:03:31 to 10:04:58	0.03	0.01	0.02
1	1	No. 7, heading	6,500	Blowing	2	10:05:53 to 10:07:23	0.04	0.01	0.03
1	1	No. 7, heading	6,500	Blowing	1	10:08:23 to 10:09:39	0.24	0.01	0.23
1	1	No. 7, heading	6,500	Blowing	2	10:10:23 to 10:11:42	0.33	0.01	0.32
1	1	No. 7, heading	6,500	Blowing	1	10:12:23 to 10:13:52	0.28	0.01	0.27
1	1	No. 7, heading	6,500	Blowing	2	10:14:34 to 10:16:12	0.34	0.01	0.34
1	1	No. 7, heading	6,500	Blowing	1	10:16:54 to 10:19:02	0.16	0.01	0.15
1	1	No. 7, heading	6,500	Blowing	2	10:19:35 to 10:21:24	0.65	0.01	0.64
1	1	No. 7, heading	6,500	Blowing	1	10:22:24 to 10:23:33	0.33	0.01	0.32
1	1	No. 7, heading	6,500	Blowing	2	10:24:30 to 10:26:19	0.10	0.01	0.09
2	1	No. 6, heading	6,400	Blowing	1	11:52:08 to 11:53:10	0.29	0.01	0.29
2	1	No. 6, heading	6,400	Blowing	2	11:53:28 to 11:54:48	0.54	0.02	0.52
2	1	No. 6, heading	6,400	Blowing	1	11:56:25 to 11:57:35	0.60	0.02	0.58
2	1	No. 6, heading	6,400	Blowing	2	11:57:55 to 11:59:05	5.19	0.09	5.09
2	1	No. 6, heading	6,400	Blowing	1	12:08:53 to 12:10:03	0.04	0.01	0.03
2	1	No. 6, heading	6,400	Blowing	2	12:12:51 to 12:13:50	0.56	0.01	0.55
2	1	No. 6, heading	6,400	Blowing	2	12:17:00 to 12:18:11	0.65	0.01	0.64
2	1	No. 6, heading	6,400	Blowing	2	12:21:25 to 12:22:30	0.69	0.01	0.68
2	1	No. 6, heading	6,400	Blowing	2	12:30:45 to 12:31:29	0.62	0.01	0.61
2	1	No. 6, heading	6,400	Blowing	2	12:34:49 to 12:35:52	0.77	0.01	0.76
2	1	No. 6, heading	6,400	Blowing	2	12:39:00 to 12:40:07	0.89	0.01	0.88
2	1	No. 6, heading	6,400	Blowing	2	12:43:33 to 12:44:51	0.57	0.01	0.55
2	1	No. 6, heading	6,400	Blowing	1	12:45:10 to 12:46:44	1.97	0.01	1.96
2	1	No. 6, heading	6,400	Blowing	2	12:48:30 to 12:49:40	1.78	0.01	1.77
2	1	No. 6, heading	6,400	Blowing	1	12:50:06 to 12:51:15	2.05	0.02	2.03
2	1	No. 6, heading	6,400	Blowing	2	12:53:15 to 12:54:35	1.02	0.01	1.00
2	1	No. 6, heading	6,400	Blowing	1	12:54:55 to 12:56:30	3.55	0.02	3.53
2	1	No. 6, heading	6,400	Blowing	2	12:58:02 to 12:59:18	3.00	0.02	2.98
3	1	No. 5, heading	7,100	Blowing	2	14:21:55 to 14:23:00	0.75	0.03	0.72
3	1	No. 5, heading	7,100	Blowing	1	14:23:22 to 14:24:43	2.03	0.07	1.96
3	1	No. 5, heading	7,100	Blowing	2	14:25:25 to 14:26:55	3.98	0.07	3.91
3	1	No. 5, heading	7,100	Blowing	1	14:27:23 to 14:28:48	5.46	0.05	5.41
3	1	No. 5, heading	7,100	Blowing	2	14:29:42 to 14:31:07	2.37	0.05	2.32
3	1	No. 5, heading	7,100	Blowing	1	14:31:35 to 14:33:04	2.66	0.04	2.62
3	1	No. 5, heading	7,100	Blowing	2	14:33:34 to 14:34:52	2.38	0.08	2.29
3	1	No. 5, heading	7,100	Blowing	1	14:35:13 to 14:36:23	2.39	0.06	2.33
3	1	No. 5, heading	7,100	Blowing	2	14:37:20 to 14:39:57	1.71	0.04	1.67
3	1	No. 5, heading	7,100	Blowing	1	14:40:24 to 14:41:35	2.81	0.07	2.73
3	1	No. 5, heading	7,100	Blowing	2	14:42:40 to 14:44:28	1.81	0.02	1.79
3	1	No. 5, heading	7,100	Blowing	1	14:44:48 to 14:45:55	2.32	0.02	2.30
3	1	No. 5, heading	7,100	Blowing	2	14:47:03 to 14:48:15	1.97	0.02	1.95
3	1	No. 5, heading	7,100	Blowing	1	14:48:50 to 14:50:08	5.20	0.01	5.18
3	1	No. 5, heading	7,100	Blowing	2	14:50:45 to 14:52:29	3.47	0.01	3.46
3	1	No. 5, heading	7,100	Blowing	2	14:54:55 to 14:56:14	1.73	0.01	1.72
3	1	No. 5, heading	7,100	Blowing	1	14:57:12 to 14:58:30	3.36	0.01	3.35

Table F-2. Shuttle car cab dust levels when loading at the face on Day 2
Note: Shaded areas show deep-cut depth, and clear areas show regular-cut depth.

Cut No.	Day	Cut Sequence	Curtain Airflow	Face Ventilation	Car No.	Time at Face	Shuttle Car Dust (mg/m³)	Intake Dust Level (mg/m³)	Adjusted Shuttle Car Dust (mg/m³)
4	2	No. 7, heading	6,500	Blowing	2	8:53:50 to 8:54:45	3.89	0.25	3.64
4	2	No. 7, heading	6,500	Blowing	1	9:01:12 to 9:02:05	1.11	0.00	1.11
4	2	No. 7, heading	6,500	Blowing	2	9:04:20 to 9:05:17	3.03	0.00	3.02
4	2	No. 7, heading	6,500	Blowing	1	9:06:39 to 9:07:40	2.09	0.01	2.09
4	2	No. 7, heading	6,500	Blowing	2	9:10:12 to 9:11:34	1.44	0.01	1.43
4	2	No. 7, heading	6,500	Blowing	1	9:12:25 to 9:14:22	2.30	0.01	2.29
4	2	No. 7, heading	6,500	Blowing	2	9:15:06 to 9:16:09	1.53	0.01	1.52
4	2	No. 7, heading	6,500	Blowing	1	9:17:22 to 9:18:22	1.47	0.01	1.46
4	2	No. 7, heading	6,500	Blowing	2	9:20:04 to 9:21:09	1.16	0.00	1.16
4	2	No. 7, heading	6,500	Blowing	1	9:21:54 to 9:22:54	3.18	0.00	3.18
4	2	No. 7, heading	6,500	Blowing	2	9:24:30 to 9:25:28	3.12	0.00	3.12
4	2	No. 7, heading	6,500	Blowing	1	9:26:17 to 9:27:10	2.33	0.00	2.32
4	2	No. 7, heading	6,500	Blowing	2	9:28:49 to 9:30:02	1.73	0.00	1.72
4	2	No. 7, heading	6,500	Blowing	1	9:30:57 to 9:32:13	2.14	0.00	2.14
4	2	No. 7, heading	6,500	Blowing	2	9:33:30 to 9:34:45	1.72	0.00	1.72
4	2	No. 7, heading	6,500	Blowing	1	9:35:25 to 9:37:30	1.47	0.00	1.46
4	2	No. 7, heading	6,500	Blowing	2	9:38:40 to 9:39:50	2.96	0.00	2.96
4	2	No. 7, heading	6,500	Blowing	1	9:40:43 to 9:42:26	2.46	0.00	2.46
5	2	No. 6, heading	7,100	Blowing	1	11:56:33 to 11:57:33	0.91	0.01	0.90
5	2	No. 6, heading	7,100	Blowing	2	12:00:35 to 12:01:51	4.27	0.02	4.25
5	2	No. 6, heading	7,100	Blowing	1	12:02:30 to 12:03:58	7.35	0.06	7.29
5	2	No. 6, heading	7,100	Blowing	2	12:04:58 to 12:07:09	3.07	0.01	3.05
5	2	No. 6, heading	7,100	Blowing	1	12:08:04 to 12:09:27	2.51	0.01	2.51
5	2	No. 6, heading	7,100	Blowing	2	12:10:54 to 12:12:00	1.01	0.01	1.00
5	2	No. 6, heading	7,100	Blowing	1	12:12:50 to 12:13:55	1.58	0.01	1.57
5	2	No. 6, heading	7,100	Blowing	2	12:16:49 to 12:18:00	1.15	0.01	1.14
5	2	No. 6, heading	7,100	Blowing	1	12:18:51 to 12:19:48	1.80	0.01	1.79
5	2	No. 6, heading	7,100	Blowing	2	12:22:00 to 12:23:18	0.99	0.02	0.98
5	2	No. 6, heading	7,100	Blowing	1	12:24:15 to 12:25:15	1.91	0.01	1.90
5	2	No. 6, heading	7,100	Blowing	2	12:27:26 to 12:28:41	1.51	0.01	1.50
5	2	No. 6, heading	7,100	Blowing	1	12:29:41 to 12:30:43	3.00	0.01	2.99
5	2	No. 6, heading	7,100	Blowing	2	12:32:40 to 12:33:53	1.43	0.01	1.42
5	2	No. 6, heading	7,100	Blowing	1	12:34:53 to 12:36:01	3.72	0.01	3.72
6	2	No. 5, heading	6,400	Blowing	1	13:46:38 to 13:47:47	0.52	0.23	0.30
6	2	No. 5, heading	6,400	Blowing	2	13:48:06 to 13:49:46	1.86	0.14	1.72
6	2	No. 5, heading	6,400	Blowing	2	13:52:54 to 13:54:40	2.47	0.15	2.32
6	2	No. 5, heading	6,400	Blowing	1	13:55:02 to 13:56:15	5.05	0.11	4.95
6	2	No. 5, heading	6,400	Blowing	2	13:59:00 to 14:00:00	1.48	0.07	1.42
6	2	No. 5, heading	6,400	Blowing	1	14:00:23 to 14:02:00	2.03	0.06	1.97
6	2	No. 5, heading	6,400	Blowing	1	14:04:30 to 14:05:45	1.03	0.06	0.97
6	2	No. 5, heading	6,400	Blowing	2	14:06:00 to 14:08:00	1.87	0.08	1.79
6	2	No. 5, heading	6,400	Blowing	1	14:08:22 to 14:14:06	Invalid	Invalid	Invalid
6	2	No. 5, heading	6,400	Blowing	2	14:14:33 to 14:16:03	3.65	0.07	3.58
6	2	No. 5, heading	6,400	Blowing	1	14:16:29 to 14:17:52	3.91	0.06	3.85
6	2	No. 5, heading	6,400	Blowing	2	14:18:32 to 14:20:00	3.90	0.05	3.84
6	2	No. 5, heading	6,400	Blowing	1	14:21:43 to 14:23:01	2.97	0.05	2.91
6	2	No. 5, heading	6,400	Blowing	2	14:23:37 to 14:25:15	3.69	0.07	3.62
6	2	No. 5, heading	6,400	Blowing	1	14:25:38 to 14:27:15	3.57	0.06	3.51
6	2	No. 5, heading	6,400	Blowing	2	14:27:51 to 14:29:49	2.20	0.05	2.15
6	2	No. 5, heading	6,400	Blowing	1	14:30:11 to 14:31:36	4.03	0.06	3.97

Table F-3. Shuttle car cab dust levels when loading at the face on Day 3

Note: Shaded areas show deep-cut depth, and clear areas show regular-cut depth.

Cut No.	Day	Cut Sequence	Curtain Airflow	Face Ventilation	Car No.	Time at Face	Shuttle Car Dust (mg/m³)	Intake Dust Level (mg/m³)	Adjusted Shuttle Car Dust (mg/m³)
7	3	No. 4, heading	6,800	Blowing	2	8:51:11 to 8:52:11	0.73	0.05	0.69
7	3	No. 4, heading	6,800	Blowing	3	8:53:28 to 8:54:00	No Data	0.06	No Data
7	3	No. 4, heading	6,800	Blowing	2	8:55:28 to 8:56:20	1.11	0.08	1.02
7	3	No. 4, heading	6,800	Blowing	3	8:57:00 to 8:57:50	No Data	0.10	No Data
7	3	No. 4, heading	6,800	Blowing	2	8:58:22 to 8:59:06	5.53	0.08	5.45
7	3	No. 4, heading	6,800	Blowing	3	9:00:13 to 9:01:41	No Data	0.07	No Data
7	3	No. 4, heading	6,800	Blowing	2	9:01:54 to 9:02:53	2.62	0.05	2.56
7	3	No. 4, heading	6,800	Blowing	3	9:03:37 to 9:04:44	No Data	0.07	No Data
7	3	No. 4, heading	6,800	Blowing	2	9:05:00 to 9:06:11	2.21	0.07	2.14
7	3	No. 4, heading	6,800	Blowing	3	9:07:10 to 9:08:15	No Data	0.05	No Data
7	3	No. 4, heading	6,800	Blowing	2	9:08:34 to 9:09:30	4.11	0.04	4.07
7	3	No. 4, heading	6,800	Blowing	3	9:10:21 to 9:11:22	No Data	0.03	No Data
7	3	No. 4, heading	6,800	Blowing	2	9:11:40 to 9:12:30	2.39	0.04	2.35
7	3	No. 4, heading	6,800	Blowing	3	9:13:28 to 9:14:28	No Data	0.03	No Data
7	3	No. 4, heading	6,800	Blowing	2	9:14:45 to 9:15:52	2.64	0.03	2.61
7	3	No. 4, heading	6,800	Blowing	3	9:16:36 to 9:17:32	No Data	0.03	No Data
7	3	No. 4, heading	6,800	Blowing	2	9:18:47 to 9:19:42	1.86	0.02	1.84
7	3	No. 4, heading	6,800	Blowing	3	9:20:05 to 9:21:17	No Data	0.02	No Data
8	3	No. 7, heading	6,400	Blowing	2	10:23:13 to 10:24:13	0.15	0.00	0.15
8	3	No. 7, heading	6,400	Blowing	1	10:25:45 to 10:26:40	0.61	0.00	0.60
8	3	No. 7, heading	6,400	Blowing	2	10:27:28 to 10:28:28	3.43	0.00	3.43
8	3	No. 7, heading	6,400	Blowing	1	10:29:30 to 10:30:34	1.61	0.00	1.61
8	3	No. 7, heading	6,400	Blowing	2	10:32:21 to 10:33:37	0.97	0.00	0.97
8	3	No. 7, heading	6,400	Blowing	1	10:34:33 to 10:35:22	1.44	0.00	1.44
8	3	No. 7, heading	6,400	Blowing	2	10:36:41 to 10:38:02	1.82	0.00	1.82
8	3	No. 7, heading	6,400	Blowing	1	10:39:04 to 10:40:30	1.50	0.00	1.49
8	3	No. 7, heading	6,400	Blowing	2	10:42:43 to 10:43:54	0.90	0.00	0.90
8	3	No. 7, heading	6,400	Blowing	1	10:45:00 to 10:46:34	3.57	0.00	3.57
8	3	No. 7, heading	6,400	Blowing	2	10:47:21 to 10:48:37	3.08	0.00	3.08
8	3	No. 7, heading	6,400	Blowing	1	10:49:34 to 10:50:44	2.33	0.00	2.32
8	3	No. 7, heading	6,400	Blowing	2	10:52:36 to 10:53:44	2.18	0.00	2.18
8	3	No. 7, heading	6,400	Blowing	1	10:54:54 to 10:56:15	2.20	0.00	2.20
8	3	No. 7, heading	6,400	Blowing	2	10:57:07 to 10:58:26	1.20	0.00	1.20
8	3	No. 7, heading	6,400	Blowing	1	11:00:14 to 11:01:40	1.01	0.00	1.01
8	3	No. 7, heading	6,400	Blowing	2	12:14:37 to 12:15:36	3.06	0.01	3.05
8	3	No. 7, heading	6,400	Blowing	1	12:16:20 to 12:17:32	1.53	0.01	1.52

Table F-4. Miner-generated dust levels during the deep- and regular-cut depth
Note: Dust levels are adjusted for productivity.

Cut No.	Cut Sequence	Depth	Curtain Airflow (cfm)	Face Ventilation	Starting Scrubber Airflow (cfm)	Ending Scrubber Airflow (cfm)	Miner Intake Dust Level (mg/m^3)	Miner Return Dust Level (mg/m^3)	Miner Generated (mg/m^3)	Cars Per (mg/m^3)	Miner Generated Adjusted (mg/m^3)
1	No. 7, heading	regular	6,500	Blowing	8,500	N/A	0.01	1.04	1.03	0.42	1.05
1	No. 7, heading	deep	6,500	Blowing	N/A	8,200	0.01	1.16	1.15	0.43	1.15
2	No. 6, heading	regular	6,400	Blowing	8,800	N/A	0.02	0.55	0.53	0.23	0.99
2	No. 6, heading	deep	6,400	Blowing	N/A	8,200	0.02	1.06	1.04	0.42	1.07
3	No. 5, heading	regular	7,100	Blowing	9,100	N/A	0.05	0.92	0.87	0.49	0.76
3	No. 5, heading	deep	7,100	Blowing	N/A	8,500	0.01	0.81	0.80	0.44	0.78
4	No. 7, heading	regular	6,500	Blowing	8,800	N/A	0.03	0.21	0.18	0.36	0.21
4	No. 7, heading	deep	6,500	Blowing	N/A	8,800	0.00	0.18	0.17	0.44	0.17
5	No. 6, heading	regular	7,100	Blowing	9,100	N/A	0.02	0.50	0.48	0.39	0.53
5	No. 6, heading	deep	7,100	Blowing	N/A	8,500	0.01	0.30	0.29	0.43	0.29
6	No. 5, heading	regular	6,400	Blowing	8,800	N/A	0.10	0.99	0.89	0.35	1.09
6	No. 5, heading	deep	6,400	Blowing	N/A	8,800	0.06	0.96	0.90	0.46	0.84
7	No. 4, heading	regular	6,800	Blowing	8,700	N/A	0.07	1.44	1.38	0.59	1.00
7	No. 4, heading	deep	6,800	Blowing	N/A	8,300	0.04	1.35	1.31	0.62	0.91
8	No. 7, heading	regular	6,400	Blowing	9,100	N/A	0.00	0.60	0.60	0.44	0.58
8	No. 7, heading	deep	6,400	Blowing	N/A	8,200	0.00	0.34	0.34	0.50	0.29

Table F-5. Miner operator dust levels during the deep- and regular-cut depths
Note: Dust levels are adjusted for productivity.

Cut No.	Cut Sequence	Depth	Curtain Airflow (cfm)	Face Ventilation	Starting Scrubber Airflow (cfm)	Ending Scrubber Airflow (cfm)	Miner Operator Exposure mg/m3	Cars Per Minute	Adj. Miner Operator Exposure mg/m3
4	No. 7, heading	regular	6,500	Blowing	8,800	N/A	3.10	0.36	3.70
4	No. 7, heading	deep	6,500	Blowing	N/A	8,800	4.28	0.44	4.18
5	No. 6, heading	regular	7,100	Blowing	9,100	N/A	3.71	0.39	4.09
5	No. 6, heading	deep	7,100	Blowing	N/A	8,500	3.22	0.43	3.22
6	No. 5, heading	regular	6,400	Blowing	8,800	N/A	4.03	0.35	4.95
6	No. 5, heading	deep	6,400	Blowing	N/A	8,800	5.19	0.46	4.85
7	No. 4, heading	regular	6,800	Blowing	8,700	N/A	0.36	0.59	0.26
7	No. 4, heading	deep	6,800	Blowing	N/A	8,300	0.44	0.62	0.31
8	No. 7, heading	regular	6,400	Blowing	9,100	N/A	0.14	0.44	0.14
8	No. 7, heading	deep	6,400	Blowing	N/A	8,200	0.22	0.50	0.19

Table F-6. Bolter-generated dust levels and operator exposures for each bolting sequence

Room No.	Entry	Depth	Position With Respect To Miner	Curtain Airflow (cfm)	Face Ventilation	Bolter Suction Left/Right (in Hg)	Left-Side Bolter Dust Levels (mg/m^3)	Right-Side Bolter Dust Levels (mg/m^3)	Upwind Bolter Dust Levels (mg/m^3)	Downwind Bolter Dust Levels (mg/m^3)	Bolter Generated Dust (mg/m^3)
1	No. 6	Regular	Upwind	7,400	Blowing	15/14	No Data	No Data	0.03	0.21	0.18
1	No. 6	Deep	Upwind	7,400	Blowing	15/14	No Data	No Data	0.02	0.19	0.17
2	No. 7	Regular	Upwind	5,500	Blowing	13/12	2.20	0.53	0.02	Invalid	Invalid
2	No. 7	Deep	Upwind	5,500	Blowing	13/12	0.30	0.35	0.01	Invalid	Invalid
3	X-cut 6-7	Regular	Upwind	Minimal	Blowing	12/12	0.53	0.50	0.04	0.40	0.36
3	X-cut 6-7	Deep	Upwind	Minimal	Blowing	12/12	0.00	0.70	0.08	0.36	0.28
4	No. 6	Regular	Upwind	None	None	12/12	1.70	1.13	0.02	0.35	0.33
4	No. 6	Deep	Upwind	None	None	12/12	0.75	0.75	0.02	0.39	0.37
5	No. 6	Regular	Upwind	10,300	Blowing	13/14	0.85	0.51	0.01	0.19	0.18
5	No. 6	Deep	Upwind	10,300	Blowing	13/14	2.16	2.02	0.00	0.47	0.46
6	No. 5	Regular	Downwind	5,900	Blowing	14/12	0.80	0.66	0.25	0.41	0.16
6	No. 5	Deep	Downwind	5,900	Blowing	14/12	5.82	1.31	0.39	0.75	0.36
7	No. 4	Regular	Upwind	4,900	Blowing	14/13	0.63	1.47	0.08	0.50	0.42
7	No. 4	Deep	Upwind	4,900	Blowing	14/13	1.99	5.49	0.02	1.01	0.98
8	No. 3	Regular	Upwind	9,100	Blowing	14/13	0.43	0.30	0.02	0.39	0.37
8	No. 3	Deep	Upwind	9,100	Blowing	14/13	0.77	0.25	0.08	0.30	0.22

www.ingramcontent.com/pod-product-compliance
Lightning Source LLC
Chambersburg PA
CBHW080306180526
45167CB00006B/2699